Engineering Materials

Zhen Fang

Rapid Production of Micro- and Nano-particles Using Supercritical Water

With 84 Figures and 2 Tables

 Springer

Zhen Fang
Chinese Academy of Science
Xishuangbanna Tropical
Botanical Garden, Biomass Group
Xuefulu 88
650223 Kunming
China, People's Republic
zhenfang@xtbg.ac.cn
zhen.fang@mail.mcgil.ca
http://brg.groups.xtbg.cn

ISBN 978-3-642-26425-2   ISBN 978-3-642-12987-2 (eBook)
DOI 10.1007/978-3-642-12987-2
Springer Heidelberg Dordrecht London New York

Cover design: WMXDesign GmbH, Heidelberg

Printed on acid-free paper

Springer is part of Springer Science+Business Media (www.springer.com)

To
my wife: CHEN, Jingmei
my parents: FANG, hong & YU, Shubi
and
my brothers: FANG, Jing & FANG, Chun

Preface

This book is talking about how to use supercritical water (SCW) to rapidly produce micro- and nano-particles of metal oxides, inorganic salts, metals and organics. It covers basic principles, experimental methodologies and reactors, particle production, characterizations and applications as well as the recent advancement. Fine particles can be produced by both chemical and physical precipitation of products from SCW. They can be used as catalysts, materials in ceramics and electronic devices and composite materials. Particles are easily produced continuously in a flow reactor in short reaction times (0.4 s~2 min) but can also be synthesized in batch reactors for long reaction times (e.g., 12 h). They can be easily studied in-situ microscopically (optical/IR/Raman/SR-XRD) in an optical micro-reactor, diamond anvil cell. The size, size distribution, crystal growth & structure, and morphology of particles can be controlled by changing the concentrations of stating materials, pH, pressures, temperatures, heating & cooling rates, organic modifications, reducing or oxidizing atmospheres, flow rates and reaction times.

This is the first book to systematically introduce using SCW for production of fine particles. It is an ideal reference book for engineers, researchers and graduate students in material science and engineering.

Acknowledgments

I would like to thank Drs. T. Ogi & T. Minowa (Biomass Technology Research Center, National Institute of Advanced Industrial Science and Technology, Japan), and Profs. K. Arai, H. Inomata, R. L. Smith Jr. and T. Adschiri (Chemical Engineering, Tohoku University, Japan), who initially introduced the hydrothermal and supercritical fluids areas to me when I worked in Japan from 1996 to1999.

Thanks are also due to Profs. J. A. Kozinski, R. I. L. Guthrie (Materials Engineering, McGill University, Canada) and I. S. Butler (Chemistry, McGill) for their guidance in my work on hydrothermal process during my work in Canada from 1999 to 2007.

Profs. W. Bassett (Geological Sciences, Cornell University) and D. Baker (Earth & Planetary Sciences, McGill) for instructions regarding DAC, Dr. I-Ming Chou (U.S. Geological Survey) for useful discussions of the pressure calculation procedure.

Drs. M. Watanabe and T. Sato (Research Center of Supercritical Fluid Technology, Tohoku University, Japan) for discussions about the experimental set-up of the batch and flow reactors.

Drs. S. Xu, H. Assaaoudi, R. Hashaikeh and A.Sobhy, who worked with me at McGill in Canada.

Contents

List of Figures

List of Tables

Nomenclature

D	diffusion coefficient	(cm^2/s)
k	reaction rate constant	$(1/s)$
K_w	ionic product for water ($[H^+][OH^-]$)	$mol^2 \cdot dm^{-6}$
P	pressure	MPa; kbar
Q	flow rate	mL/min
R	gas constant; ruby fluorescence R-lines	$J \cdot K^{-1} \cdot mol^{-1}$
T	temperature	K
V	volume	m^3; mL
x	mole fraction	
X	conversion degree	
y_2	solubility of solute	mol%; wt%
ε	dielectric constant	
γ	activity coefficient	
η	viscosity	cP
κ_T	isothermal compressibility at T	1/Mpa
λ	wave length; thermal conductivity	nm; w/(m·k)
ρ	density	kg/m^3; mol/mL
τ	reaction time	Min; s
υ	specific volume	m^3/kg
ϖ	Raman wavenumber of pressure sensor	cm^{-1}
ω	constant determined by a reaction system (> 0)	

Superscripts

G, g   gas
S, s   solid
∞ infinite dilute mixture

Subscripts

0	pure component, initial
1	ruby fluorescence R_1 line
2	solute; component 2; ruby fluorescence R_2 line
c	critical (point; curve)
h	homogenization
m	mole; melting
r	reaction
sc	supercritical
Ac	acetate

Abbreviations

CC	critical curve
CNT	carbon nanotubes
CP	critical point
DAC	diamond anvil cell
DLS	dynamic light scattering
DSC	differential scanning calorimeter
DTA	differential thermal analysis
EDX	energy dispersive X-ray spectrometry
EOSW	equation of state of water
EPMA	superprobe electron probe micro-analyzer
FBT	fluidized bed temperature
FT-IR	Fourier transform infrared
g	gas
GC-MS	gas chromatography-mass spectrometry
GC-TCD	gas chromatography-thermal conductivity detector
HB	hydrogen bonding
HPLC	high-pressure liquid chromatography
IC	ion chromatography
ICP	inductively coupled plasma spectrometry
KTO	potassium hexatitanate ($K_2Ti_6O_{13}$)
L(1)	NO_3^-, CH_3COO^- or other anions; liquid
llg	liquid-liquid-gas
L$-$V(1-v)	liquid-vapor
Me	metal
PET	polyethylene terephthalate
PGSS	particles from gas-saturated solutions
PVA	polyvinyl alcohol
RESS	rapid expansion of supercritical solution
SAS	supercritical antisolvent
SCF	supercritical fluid
SCW	supercritical water

SCWO	SCW oxidation
SEM	scanning electron microscope
SR-XRD	synchrotron radiation XRD
SS	stainless steel
SSR	solid-state reaction
SubCW	subcritical water
TEM	transmission electron microscope
TGA	thermogravimetric analysis
TPA	terephthalic acid
XRD	X-ray diffraction
YAG	aluminium yttrium garnet

Chapter 1
Introduction

Abstract A supercritical fluid (SCF) is any substance at a temperature and pressure above its critical point (CP). SCF has properties between those of a gas and a liquid, and has a strong solubility that changes significantly near its CP. Small changes in pressure or temperature result in large variation in solubility, allowing production of fine particles via precipitation of a solute from supercritical solution by rapidly exceeding the saturation point of the solute by dilution, antisolvent, depressurization and attemperation. In addition to the above properties for SCFs, supercritical water (SCW) has its peculiar properties due to the hydrogen bonding (HB) among water molecules. In SCW, more than 70% HB is destroyed to a non-polar dimer, and further decomposed to monomer, $[H^+]$ and $[OH^-]$ ions. Therefore, SCW becomes a weakly-polar solvent that can dissolve non-polar organics to produce fine particles via precipitation. On the other hand, high ion concentration of $[H^+]$ and $[OH^-]$ provides both basic and acidic reaction conditions for fine particles synthesis.

1.1 Background

A fluid is called supercritical when its temperature (T) and pressure (P) are higher than its critical point (CP; $> T_c$ and P_c). A supercritical fluid (SCF) is a single phase that possesses the characteristics of both liquid and gaseous substances. There is no surface tension in SCF, as there is no liquid/gas phase boundary. For pure substances, there is an inflection point in the critical isotherm on a pressure-volume (P–V) diagram. This means that at the CP: $(\partial P/\partial V)_T = (\partial^2 P/\partial V^2)_T = (\partial^2 P/\partial T^2)_V = 0$. Fluids near their CP have dissolving power comparable to that of liquids, are much more compressible than dilute gases, and have transport properties intermediate between gas-like and liquid-like (see Table 1.1) [1]. They have sufficient density to give appreciable dissolving power; but the diffusivity of solutes in SCFs is higher than in liquids, and the viscosity of SCFs is lower, facilitating mass transport. This unusual combination of physical properties can be advantageously exploited in new kinds of materials processing. In Table 1.2, the critical properties

Z. Fang, *Rapid Production of Micro- and Nano-particles Using Supercritical Water*,
Engineering Materials 30, DOI 10.1007/978-3-642-12987-2_1,
© Springer-Verlag Berlin Heidelberg 2010

Table 1.1 Physical and transport properties of gases, liquids and supercritical fluids

Property	Gas	Supercritical fluid	Liquid
Density, ρ (kg/m^3)	1	100~800	1,000
Viscosity, η (cP)	0.01	0.05~0.1	0.5~1.00
Diffusion coefficient, D (cm^2/s)	1~10	0.01~0.1	0.001
Thermal conductivity, λ [w/(m$\bullet$k)]	0.004~0.03	0.02~0.08	0.08~0.25

Table 1.2 Critical conditions for various supercritical solvents

Fluid	Critical temperature T_c (K)	Critical pressure P_c (MPa)	Critical density ρ_c (kg/m^3)
Carbon dioxide (CO_2)	304.1	7.4	469
Water (H_2O)	647.1	22.1	322
Methane (CH_4)	190.4	4.6	162
Ethylene (C_2H_4)	282.4	5.0	215
Ethane (C_2H_6)	305.3	4.9	203
Propylene (C_3H_6)	364.9	4.6	232
Propane (C_3H_8)	369.8	4.3	217
Methanol (CH_3OH)	512.6	8.1	272
Ethanol (C_2H_5OH)	513.9	6.1	276
Acetone (C_3H_6O)	508.1	4.7	278

(T_c, P_c and critical density ρ_c) are shown for some components, which are commonly used as SCFs [2]. Carbon dioxide and water are the most frequently used in a wide range of applications due to their low cost and non-toxicity, including particle production, extraction, dry cleaning, biomass conversion and hazardous wastes treatment.

At an infinite dilute binary mixture, the solute's partial molar volume $V_{2,m}^{\infty}$ diverges at the solvent's CP, to $-\infty$ in attractive mixtures, and to $+\infty$ in repulsive ones [3]:

$$V_{2,m}^{\infty} \propto \kappa_T = -(1/V_m)(\partial V_m/\partial P)_T \rightarrow \pm\infty \tag{1.1}$$

where κ_T - solvent's compressibility; V_m – solvent volume per mole; P – pressure; T – temperature; subscript 2 – solute, m – mole; superscript ∞ – infinite. For example, the partial molar volume of naphthalene at around supercritical point of ethylene (285 K and 0.007 mol/mL) diverges to about −15,000 mL/mol from 0 at 0.01 mol/mL (CP: $T_c = 282.4$ K, $\rho_c = 0.0077$ mol/mL, $P_c = 5$ MPa) (see Fig. 1.1) [3]. The solute's solubility y_2 is calculated by the following equation [4]:

$$(\partial y_2/\partial P)_T(1/y_2 + \partial \mathrm{dln}\gamma_2^G/\partial y_2) = (V_{2,m}^S - V_{2,m}^G)/RT \tag{1.2}$$

where γ – activity coefficient; superscript S – solid, G – gas; R – gas constant.

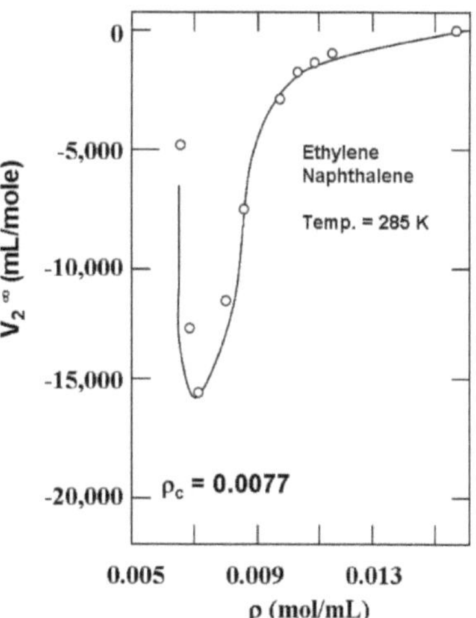

Fig. 1.1 Infinite dilution partial molar volume of naphthalene in supercritical ethylene [$T_c = 282.4$ K, $\rho_c = 0.0077$ mol/mL] at 285 K. Reprinted with permission from [3]. Copyright © 1986, American Chemical Society

Near CP, the divergence of partial molar volume ($V^G_{2,\,m}$) leads to great change of $(\partial y_2/\partial P)_T$ according to the Eq. (1.2). Therefore, varying of pressure will cause significant solubility enhancement (Fig. 1.2) [5]. Figure 1.2 also shows a sharp rise in solubility at an isobar in supercritical ethylene as temperature increases from 285 to 323 K. Experimental data showed that solubility of naphthalene in near critical region of ethylene could increase by around 1,000 times [4]. Using the method of precipitation of solute from supercritical solution with its great solubility change, several methods are used to produce micro- and nano-particles as described below.

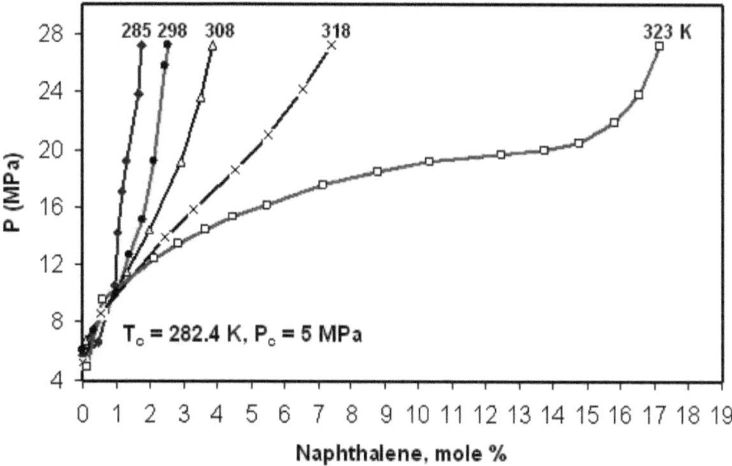

Fig. 1.2 Solubility of naphthalene in ethylene ($T_c = 282.4$ K, $P_c = 5$ MPa). Reprinted with permission from [5]. Copyright © 1953, American Chemical Society

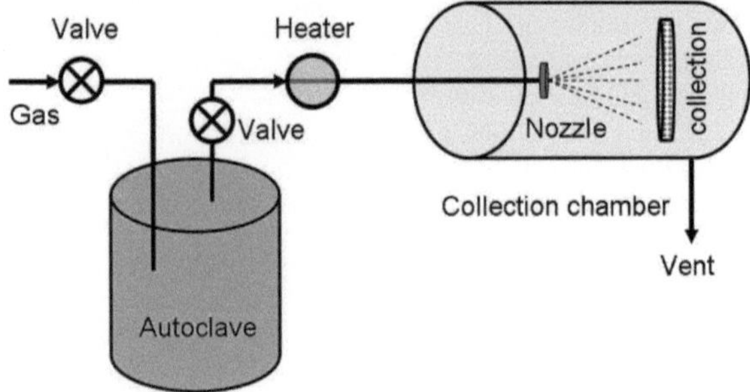

Fig. 1.3 Rapid expansion of supercritical solution (RESS) process

1.2 Rapid Expansion of Supercritical Solution (RESS) Process

The principal of the rapid expansion of supercritical solution (RESS) process (see Fig. 1.3) is that the solution of an SCF saturating with the solute, then depressurizing this solution through a heated nozzle into a low pressure chamber in order to cause an extremely rapid nucleation of the solute in form of very small particles. Microparticles are typically obtained (Nanoparticles formation is also reported), such as polymers (e.g., polyethylene), inorganic and organic materials (e.g., AgI, benzoic acid) and pharmaceutical compounds (e.g., aspirin) [6]. Details of the process can be seen in the review papers [6, 7].

1.3 Supercritical Antisolvent (SAS) Process

In this process (Fig. 1.4), an SCF is used as an anti-solvent that causes precipitation of the solute dissolved in an organic-solution. The process requires that solute is insoluble in SCF. SAS process (and its variants) has been used to re-crystallize many different products such as explosives, polymers, pharmaceuticals, coloring materials, etc. [6]. Details can be referred to the review papers [6–8].

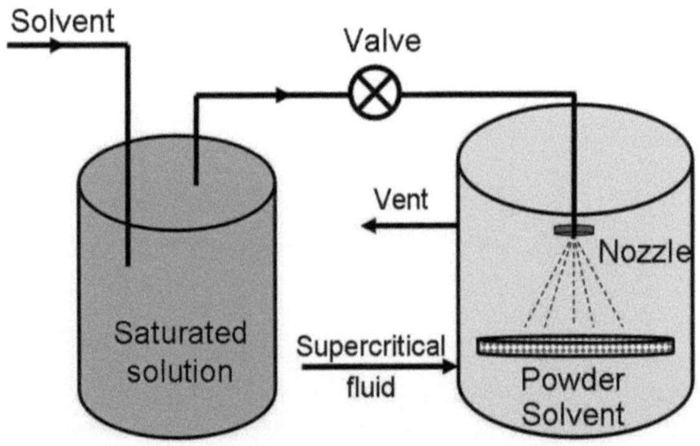

Fig. 1.4 Supercritical antisolvent (SAS) process

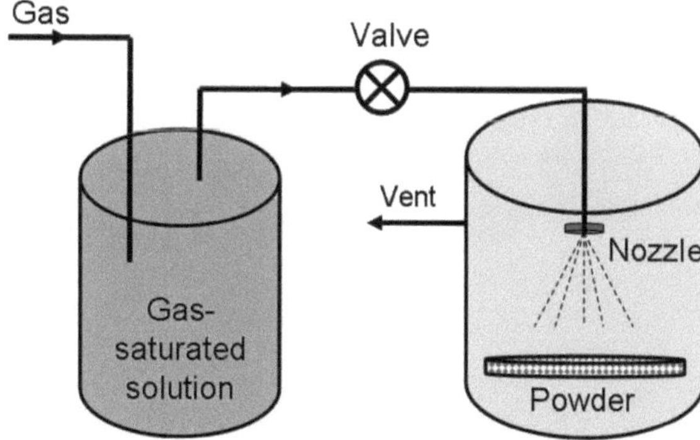

Fig. 1.5 Particles from gas-saturated solutions PGSS process

1.4 Other Physical Processes

SAS re-crystallization: This process is mostly used for re-crystallization of solid dissolved in a solvent with the aim of obtaining either small size particles or large crystals, depending on the growth rate controlled by the anti-solvent pressure variation rate [6, 7].

PGSS (Particles from Gas-Saturated Solutions): In this process (Fig. 1.5), an SCF dissolving into a liquid substrate, or a solution of the substrate(s) in a solvent, or a suspension of the substrate(s) in a solvent and followed by a rapid depressurization of this mixture through a nozzle causes the formation of solid particles or liquid droplets. PGSS process is used in the generation of powders from various substances such as polymers, waxes, resins and natural products.

1.5 Supercritical Water Process

In addition to the above properties for SCFs, supercritical water (SCW; $> T_c = 647$ K & $P_c = 22.1$ MPa, see Fig. 1.6, the shaded area) has its peculiar properties due to the hydrogen bonding among water molecules.

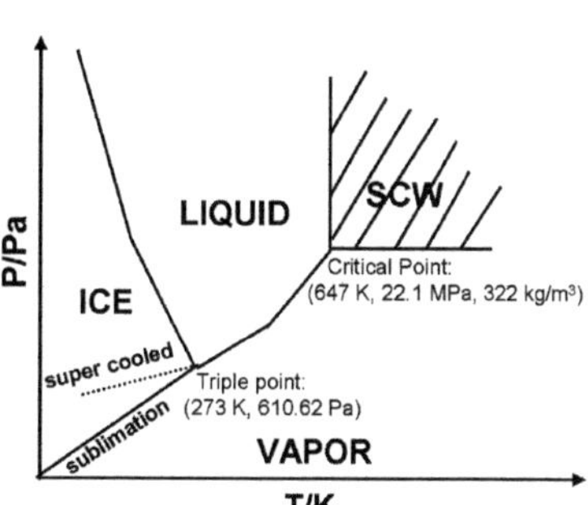

Fig. 1.6 Phase diagram of water

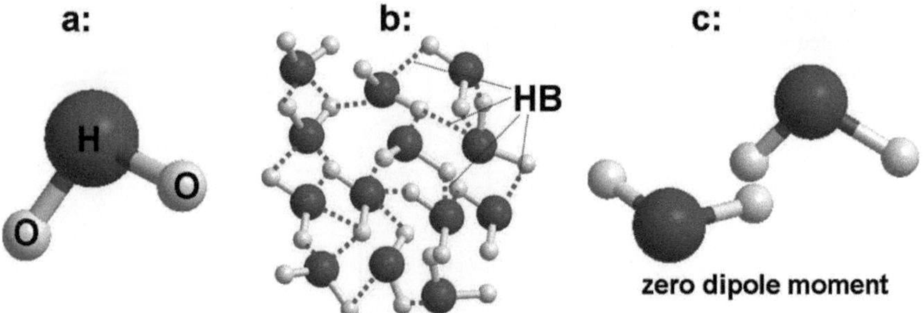

Fig. 1.7 Structure of water: (**a**) water molecule, (**b**) 3-D structure of water "polymer", and (**c**) a non-polar water dimer in SCW

Water molecules (H_2O) have four, approximately tetrahedrally arranged (H–O–H angle of 104.45°; Fig. 1.7a), sp^3-hybridized electron pairs, two of which are associated with hydrogen atoms leaving the two remaining lone pairs. More electronegative oxygen than hydrogen and the bond angle make water a polar molecule [9]. In a water molecule, the electronegative atom oxygen effectively strips hydrogen of its only electron, leaving a nearly unshielded proton exposed. This proton is powerfully attracted to the oxygen atom on a neighboring molecule that has unshared electron pairs. The proton and the lone pair form a bond called hydrogen bonding (HB) [10]. Thus, water forms a 3-D hydrogen-bonded network or clustering structure, as "poly-water [$(H_2O)_n$]" (Fig. 1.7b). Its high- or low dielectric constant (ε) is also associated with the distortion or breaking down of hydrogen bonds [11, 12].

The polar characteristics and network structure make water a good solvent for ionic salts or polar compounds but with little solubility for non-polar molecules. However, the properties of water change dramatically above CP. It is found that only 29~66% HB of water at ambient condition (298 K and 0.1 MPa) remains in SCW at 673 K and 450~1,100 kg/m^3 [13–19]. The destruction of HB substantially reduces dielectric constant [20]. Near CP, hydrogen-bonded network of water breaks up to dimers in majority, part of the dimers further to monomer [13–16]. Monomer may breakdown further and leads to the evolution of *[H$^+$]* and *[OH$^-$]* ions [15–17]. The dimers may have a point of symmetry and a zero dipole moment (Fig. 1.7c) [21].

Therefore, SCW becomes a weakly-polar solvent with both acidic and basic characters: (i). The dielectric constant (ε) of SCW (Fig. 1.8) is in the range of from 2 to 30 (compared to 78 at 298 K and 0.1 MP) [22], which means it becomes weakly-polar and possible to dissolve non-polar organic substances to provide a homogenous phase for reactions (or precipitation of solute for production of fine particles) but has low solubility for ionic inorganic salts; (ii). The ion product of SCW $\{K_w = [H^+] \bullet [OH^-]\}$ increases considerably (e.g., at 673 K and 50 MP, $K_w = 10^{-11.88}$ or *[H$^+$]* $= [OH^-] = 11.5$ times of those in water at 298 K and 0.1 MP; Fig. 1.9) [23]). High K_w {or *[OH$^-$]*} can promote metal salts hydrolyze to hydroxides without adding any base catalysts.

Fig. 1.8 Dielectric constant
(ε) of water

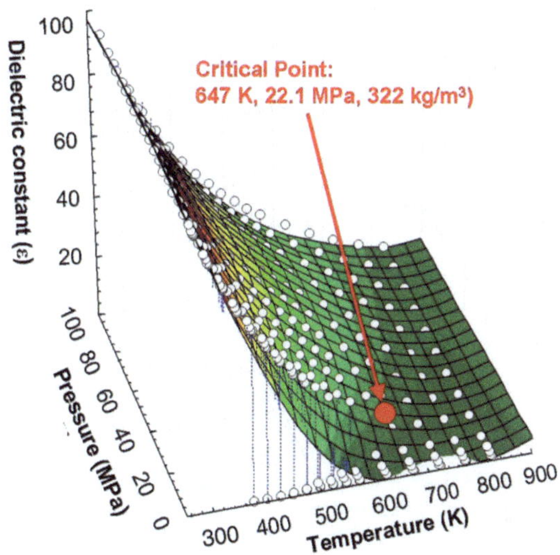

Fig. 1.9 The negative
Logarithm of the ion products
of water ($-\mathrm{Log}K_w$)

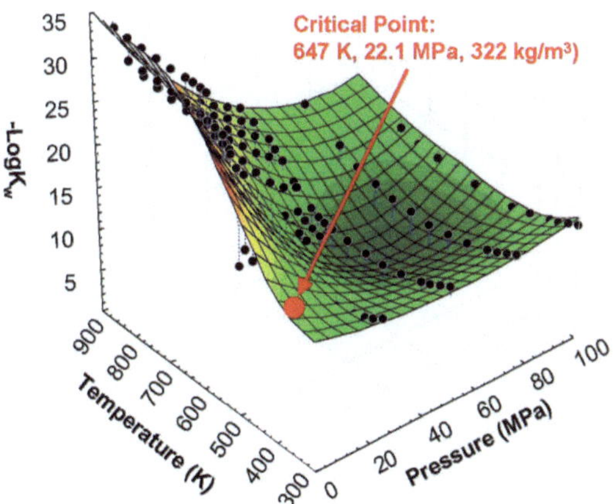

These properties (ε and K_w) make particle synthesis reactions faster in SCW
than in subcritical water (SubCW). They also decide on equilibrium constant K and
solubility of the oxide particles synthesized [24]. Fine particles can also be produced
by depressurizing or quenching the homogenous supercritical solution to precipitate
solute as particles.

The following chapters will introduce and discuss the SCW production of metal
oxides, metals, inorganic salts and organics particles, which can be used as catalysts,

sensors, recording-media battery-cathode, opto-electronic, ceramic and composite materials. Many nano-sized oxides were synthesized that can miniaturize the size of devices and increase catalytic effect {electronic, geometric and support effects [25]}. Organics particle production (such as polymers and biomass) via novel precipitation and reaction of solutes in supercritical water will also be discussed.

References

1. C.A. Eckert, B.L. Knutson, P.G. Debenedetti, Supercritical fluids as solvents for chemical and materials processing. Nature **383**, 313–318 (1996)
2. R.C. Reid, J.M. Prausnitz, B.E. Poling, *The Properties of Gases and Liquids*, 4th edn. (McGraw-Hill, New York, NY, 1987)
3. C.A. Eckert, D.H. Ziger, K.P. Johnston, S.J. Kim, Solute partial molar volumes in supercritical fluids. J. Phys. Chem. **90**, 2738–2746 (1986)
4. R.T. Kurnik, R.C. Reid, Solubility extrema in solid-liquid equilibria. AIChE J. **27**, 861–863 (1981)
5. G.A.M. Diepen, F.E.C. Scheffer, The solubility of naphthalene in supercritical ethylene. J. Phys. Chem. **57**(6), 575–577 (1953)
6. J. Jung, M. Perrut, Review – Particle design using supercritical fluids: Literature and patent survey. J. Supercrit. Fluids **20**, 179–219 (2001)
7. R. Marr, T. Gamse, Use of supercritical fluids for different processes including new development – A review. Chem. Eng. Prog. **39**, 19–28 (2000)
8. E. Reverchon, Supercritical antisolvent precipitation of micro-and nano-particles. J. Supercrit. Fluids **15**, 1–21(1999)
9. P. Atkins, L. Jones, *Chemistry: Molecules, Matter, and Change*, 3rd edn. (W. H. Freeman & Co. and Sumanas, Inc., New York, NY 1997)
10. T.L. Brown, H.E. LeMay, B.E. Bursten, *Chemistry: The Central Science*, 8th edn. (Prentice-Hall, Upper Saddle River, NJ, 1999)
11. G.A. Jeffrey, *An Introduction to Hydrogen Bonding: Ch. 8. Water, Water Dimers, Ices, Hydrates* (Oxford University Press, New York, NY, 1997)
12. D.S. Eisenberg, W. Kauzmann, *The Structure and Properties of Water* (Oxford University, New York, NY, 1969)
13. M.M. Hoffmann, M.S. Conradi, Are there hydrogen bonds in supercritical water? J. Am. Chem. Soc. **119**, 3811–3817 (1997)
14. M. Boero, K. Terakura, T. Ikeshoji, C.C. Liew, M. Parrinello, Water at supercritical conditions: A first principles study. J. Chem. Phys., **125**, 2219–2227 (2001)
15. Y. Ikushima, K. Hatakeda, N. Saito, M. Arai, An in situ Raman spectroscopy study of subcritical and supercritical water: The peculiarity of hydrogen bonding near the critical point. J. Chem. Phys. **108**, 5855–5860 (1998)
16. Y. Ikushima, K. Hatakeda, O. Sato, T. Yokoyama, M. Arai, Acceleration of synthetic organic reactions using supercritical water: Noncatalytic beckmann and pinacol rearrangements. J. Am. Chem. Soc. **122**, 1908–1918 (2000)
17. Y.E. Gorbaty, A.G. Kalinichev, Hydrogen bonding in supercritical water. 1. Experimental results. J. Phys. Chem. **99**, 5336–5340 (1995)
18. R.D. Mountain, Molecular dynamics investigation of expanded water at elevated temperatures. J. Chem. Phys. **90**, 1866–1870 (1989)
19. R.B. Gupta, C.G. Panayiotou, I.C. Sanchez, K.P. Johnston, Theory of hydrogen bonding in supercritical fluids. AIChE J. **38**, 1243–1253 (1992)
20. N. Yoshii, S. Miura, S. Okazaki, Density fluctuation and hydrogen-bonded clusters in supercritical water: A molecular dynamics analysis using a polarizable potential model. Bull. Chem. Soc. Jpn. **72**, 151–162 (1999)

21. M.P. Bassez, Is high-pressure water the cradle of life? J. Phys. Condens. Matter. **15**, L353–L361 (2003)
22. M. Uematsu, E.U. Franck, Static dielectric constant of water and steam. J. Phys. Chem. Ref. Data **9**(6), 1291–1306 (1980)
23. W.L. Marshall, E.U. Franck, Ion product of water substance, 0–1000°C, 1–10,000 bars new international formulation and its background. J. Phys. Chem. Ref. Data **10**(2), 295–304 (1981)
24. T. Adschiri, Y. Hakuta, K. Arai, Hydrothermal synthesis of metal oxide fine particles at supercritical conditions. Ind. Eng. Chem. Res. **39**, 4901–4907 (2000)
25. F. Raimondi, G.G. Scherer, R. Kotz, A. Wokaun, Nanoparticles in energy technology: Examples from electrochemistry and catalysis. Angew. Chem. Int. Ed. **44**, 2190–2209 (2005)

Chapter 2
Supercritical Water Process

Abstract Hydrothermal synthesis of particles is usually carried out at subcritical conditions in batch reactors. In supercritical region, reaction rate increases dramatically due to low dielectric constant (ε) of SCW. Therefore, fine particles (e.g., metal oxides) are rapidly synthesized (e.g., 0.4 s~2 min) in a continuous SCW process. In the oxide synthesis, an ionic metal salt is first hydrolyzed to metal hydroxide, which is then dehydrated to form metal oxide crystals by precipitating from SCW solution. Here, three types of reactors (batch, flow and diamond anvil cell) are introduced in details for the study of SCW synthesis of particles. In general, batch reactors are used for long time process, and flow reactors for short time synthesis. Diamond anvil cell is used for in-situ visual and spectroscopic study of particle formation. Continuous flow reactors allow a better control of experimental conditions (e.g., temperature, pressure, time, concentration, pH and heating rate) that lead to formation of smaller particles with uniform size.

2.1 Introduction

Conventional methods prepared for oxide particles are high temperature solid-state reaction (SSR) [1, 2], high-energy ball milling [3] sol-gel [1, 4, 5], microemulsion methods [6, 7], oxidation techniques [8] and co-precipitation with ageing of solutions [9, 10]. However, many of these processes are energy- and time-consuming (high temperature, multi-step syntheses or ageing) and environmentally unfriendly [11]. Hydrothermal (sub- or super-critical water) synthesis is another method to produce fine particles. This process has been carried out since the end of the nineteenth century, mainly in the production of synthetic minerals to imitate natural geothermal processes [12, 13]. Since then many highly crystalline metal oxides have been produced [11]. However, most of the syntheses were undergone long-time reaction in batch reactors in SubCW conditions [14, 15]. Continuous SCW synthesis is a relative novel technique to produce fine particles. Arai and Adschiri (1992, Tohoku University, Japan) pioneered the rapid one-step synthesis of metal oxide nano-crystals in SCW in continuous flow reactors [16, 17].

Z. Fang, *Rapid Production of Micro- and Nano-particles Using Supercritical Water,*
Engineering Materials 30, DOI 10.1007/978-3-642-12987-2_2,
© Springer-Verlag Berlin Heidelberg 2010

In the process of hydrothermal synthesis, low-cost precursors (e.g., nitrate or acetate salts, hydroxides) are first dissolved in water. The salt solution is introduced into a reactor operated at sub- or super-critical conditions. This method has a potential to adjust the direction of crystal growth, morphology, particle size and size distribution, because of the controllability of thermodynamics and transport properties by pressure and temperature. The synthesis reactions taking place in the reactor are considered to be as follows [17, 18]:

$$MeL_x + x\,H_2O \rightarrow Me(OH)_x + x\,HL \tag{2.1}$$

$$Me(OH)_x \rightarrow MeO_{x/2} \downarrow + x/2\,H_2O \tag{2.2}$$

$$MeO_{x/2} + x/2\,H_2 \rightarrow Me + x/2\,H_2O \tag{2.3}$$

where $L = NO_3^-$, CH_3COO^- or other anions; Me = metal.

In Eq. (2.1), metal salt is first hydrolyzed to metal hydroxide, which is then dehydrated, according to Eq. (2.2), forms metal oxide crystals with micro- or nano-size by precipitating from the solution. Metal particles can be produced by hydrogen reduction of the metal oxide particles (Eq. 2.3) [18]. In an isobar (Fig. 1.9), K_w or *[OH$^-$]* concentration increases significantly upon reaching supercritical region. Therefore, SCW promotes hydrolysis reaction (Eq. 2.1) and enhances production of oxide particle. These reactions (Eqs. 2.1, 2.2, and 2.3) also explain how to remove water-soluble heavy metal cations and recover metal oxides using hydrothermal SCW techniques [19, 20].

In our previous work [20], water-soluble heavy metal cations (100% Pb, 97.6% Cd, and 87.3% Cr) were converted to water-insoluble particles during SCW oxidation (SCWO) of organic wastes. In addition to the formation of metal oxides (Eqs. 2.1, 2.2, and 2.2), it was found that CO_2 and acetate from SCWO of organics could help to remove heavy metal cations via formation of water insoluble carbonate salts (Eq. 2.4). X-ray diffraction (XRD) spectra indicated the presence of $PbCrO_4$ and $Al_2Si_2O_5(OH)_4$ crystals in the ash (Fig. 2.1).

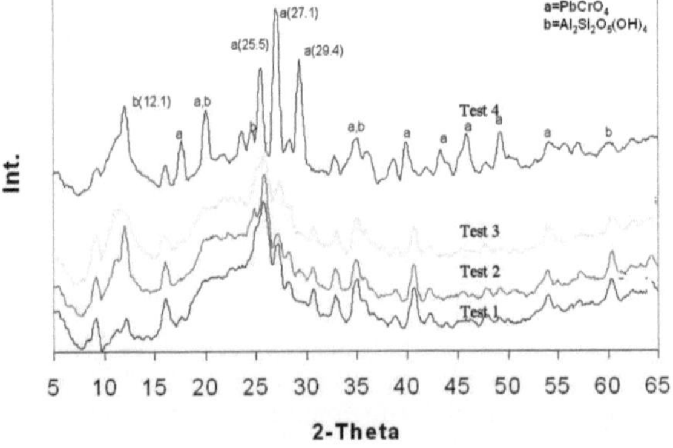

Fig. 2.1 XRD spectra of ashes from SCWO of organic wastes with heavy metal cations (tests 1–4 at different temperatures & excessive O_2). Reprinted with permission from [20]. Copyright © 2000, American Chemical Society

Super-probe electron probe micro-analyzer (EPMA) results showed the ash contained very fine and highly concentrated Pb (4.4–75.3%), and a close connection between Pb and Cr but no relation between Pb and Cd in the ash (Fig. 2.2). In most of the locations where Pb was present, Cr was also seen, but no Cd was detected (Fig. 2.2a, b, points A–E, and Fig. 2.2d, e, point A). The main solid fine particles were CdO, $CdCO_3$, CrO_2, $HCrO_2$, $PbCrO_4$, $PbCO_3$ and PbO_x. In SCWO of organics, more reactions are likely to follow:

$$MeO_{x/2} + CO_2 \rightarrow Me(CO_3)_y \downarrow \tag{2.4}$$

$$CrO_2 + O_2 \rightarrow CrO_4^{2-} \tag{2.5}$$

$$Pb^{2+} + CrO_4^{2-} \rightarrow PbCrO_4 \downarrow \tag{2.6}$$

Hydrothermal synthesis of fine crystal particles in SCW is a rapid process. Adschiri et al. [21] explained the higher synthesis reaction rate in SCW due to its low dielectric constant (ε) as compared with the rate in SubCW, according to the theory of Born:

$$\ln k = \ln k_0 - (\omega/R)(1/\varepsilon - 1/\varepsilon_0)/T \tag{2.7}$$

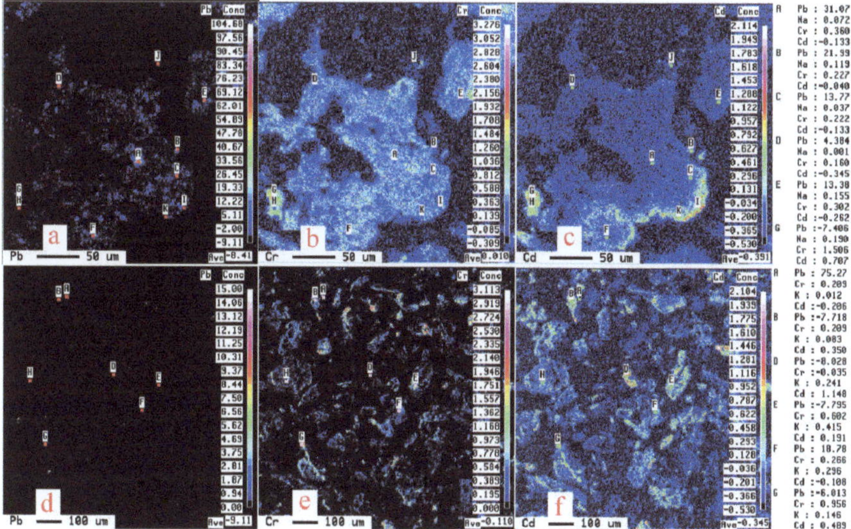

Fig. 2.2 EPMA images of analytical results for the selected ash: (i) a small particle (50 μm): **a** (Pb), **b** (Cr), **c** (Cd); (ii) a large mapping area (100 μm): **d** (Pb), **e** (Cr), **f** (Cd). Reprinted with permission from [20]. Copyright © 2000, American Chemical Society

where ω – constant determined by a reaction system (> 0); ε – dielectric constant; k – reaction rate constant; k_0 – reaction rate at constant dielectric constant ε_0 and T_0.

As temperature increases to SCW region from an initial temperature T_0, ε drops significantly (Fig. 1.8), thus $(1/\varepsilon - 1/\varepsilon_0)$ increases sharply (Fig. 2.3). So, ln k vs. 1/T [slope: $(\omega/R) \bullet (1/\varepsilon - 1/\varepsilon_0)$] rises greatly according to Eq. 2.7. Experiments were carried out to verify the Eq. 2.7 at different reaction time (τ) and temperature for the aqueous solution of $Al(NO_3)_3$ (0.01 mol/L) system at 30 MPa in a flow reactor [21]. The $[Al^{3+}]$ conversion rate, X was calculated by measuring its concentration changes in the effluent. The rate constant (k) was obtained in Fig. 2.4 from the slope of $-\ln(1 - X)$ vs. τ. Therefore, Arrhenius first order plot of ln k with 1/T was given as showed in Fig. 2.5.

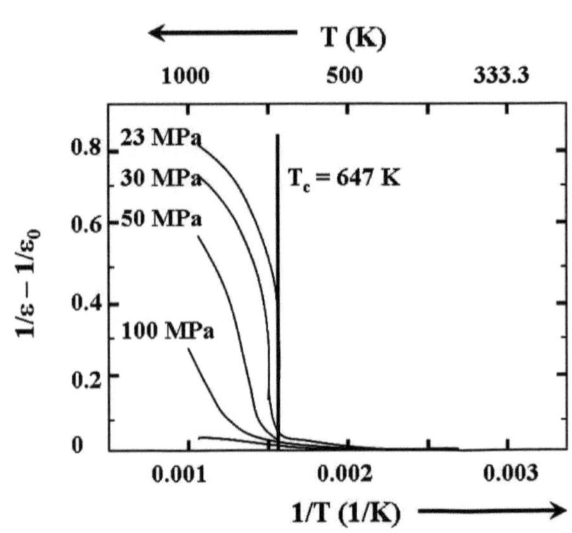

Fig. 2.3 $(1/\varepsilon - 1/\varepsilon_0)$ plotted against 1/T. Reprinted with permission from [21]. Copyright © 2000, American Chemical Society

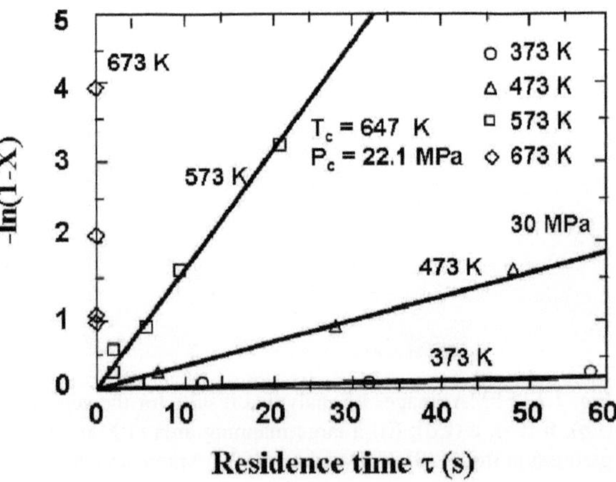

Fig. 2.4 $-\ln(1-X)$ vs. time curves at 30 MPa for $Al(NO_3)_3$ (0.01 mol/L). Reprinted with permission from [21]. Copyright © 2000, American Chemical Society

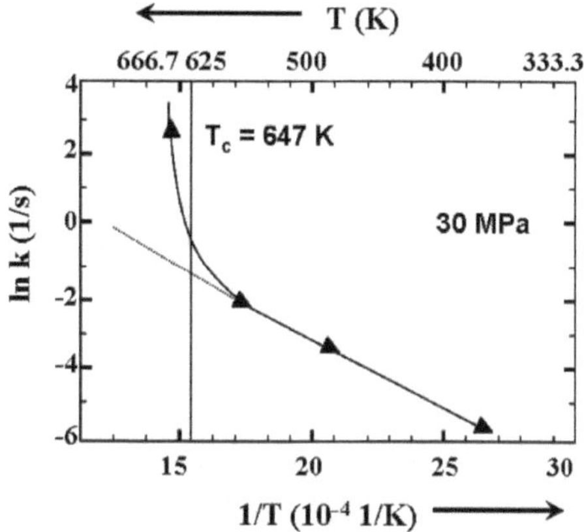

Fig. 2.5 Arrhenius plot of the apparent first-order rate constant at 30 MPa for Al(NO$_3$)$_3$. Reprinted with permission from [21]. Copyright © 2000, American Chemical Society

In the range of subcritical temperature, the rate constant fell on a straight line, but increased significantly above the critical temperature. This enhancement in rate in the critical region confirms the result predicted by Eq. (2.7). Synthesis of oxide fine particles in SCW is a rapid process (typically from 0.4 s to 2 min) in a continuous flow reactor.

It is important to have appropriate reactors to do experiments. Three types of reactors, batch, flow and diamond anvil cell (DAC) are introduced next for different applications and purposes in fine particle production.

2.2 Batch Reactor

Batch and flow reactors are widely used for the hydrothermal processes. The batch reactor could be an autoclave heated by an electric heater (Fig. 2.6; FCFD05-30, Yantai Jianbang chemical mechanical Co., Ltd. Shandong, China) or a tubular reactor heated by a salt bath (or fluidized sand bed) (Fig. 2.7). Owing to slow heating rates in autoclaves (e.g., 0.18 K/s; 30 min heated to 623 K from room temperature), smaller tubular reactors with higher heating rates heated by a fluidized sand bath (e.g., 5.4 K/s; 2 min heated to 673 K from room temperature) are widely used in SCW reactions (Fig. 2.7). The 316 stainless-steel (SS) or inconel tubular batch reactor (6 mL; length of 105 mm, OD of 12.7 μm, ID of 8.5 μm) is widely used to investigate hydrothermal reactions. The reactor is connected to a transducer (Dynisco E242) for on-line pressure and temperature (J-type thermocouple) measurements, as well as to a data acquisition system (Strawberry tree, DS-12-8-TC). Both the transducer and the reactor can be used up to 35 MPa at 811 K.

Solid or liquid samples and water are loaded into the reactor. After being sealed and connected to the data acquisition system, the reactor is submerged in a fluidized sand bath (Omega FSB-3), where temperature can be controlled up to 873 K within 1 K. The uniformity of the sand temperature is within 0.5 K. The sample is heated at a rate of 3.5 K/s up to 873 K. Figure 2.8 gives an example of temperature

Fig. 2.6 Photograph of a commercial autoclave set-up in our Lab

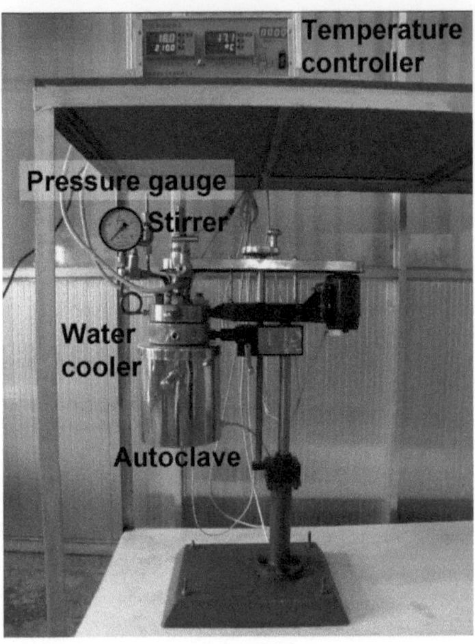

and pressure profiles measured at different fluidized bed temperature (FBT) with or without loading H_2O_2 (49 wt%, 1 mL). H_2O_2 will decompose to H_2O and O_2 (mainly at low temperature of 426 K) [22]. Usually pressure can also be calculated by an equation of state of water (EOSW) [23] by knowing temperature measured and water density ($\rho =$ water mass/reactor volume; kg/m^3).

After reactions, the reactor is quenched in cold water. The reactor is then connected to a 500-mL syringe for the collection and volume measurements of the gas phase generated during reactions. The gas is analyzed by Gas Chromatograph-Thermal Conductivity Detector (GC-TCD). The reaction mixture is separated into three phases: (1) aqueous phase (10 mL), which is obtained by washing products with water *via* 25-nm pore size membrane filter; (2) acetone or benzene phase (10 mL), which is obtained by washing the remainder with acetone or benzene; and (3) fine particles, collected on the filter paper (or separated by using a super-speed centrifuge) and dried at 373 K for 24 h. The aqueous phase is analyzed for anion and metal ion concentrations using Ion Chromatograph (IC) and Inductively Coupled Plasma spectrometry (ICP), respectively, in order to estimate the conversion rate for the fine particles synthesized. The acetone or benzene phase is analyzed by High-Pressure Liquid Chromatography (HPLC) and GC-Mass Spectrometry (GC-MS). Fine particles are analyzed their elemental distribution by an EPMA and Energy Dispersive X-Ray spectrometry (EDX), their structure is analyzed by XRD. The particles are also characterized by Dynamic Light Scattering (DLS), Mössbauer Spectroscopy, Fourier Transform Infrared (FTIR)/Raman Microscopy, Transmission Electron Microscope (TEM), Scanning Electron Microscopy (SEM), Micromeritics Surface Area and Pore Size Analyzer, Differential Scanning Calorimeter (DSC) and Thermogravimetric Analysis (TGA).

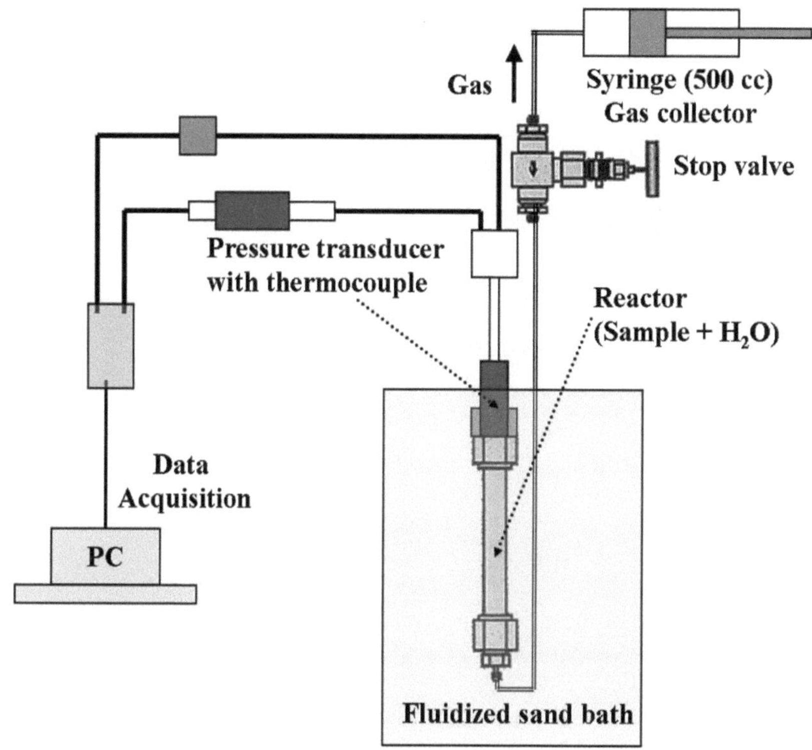

Fig. 2.7 A tubular batch reactor set-up in our lab

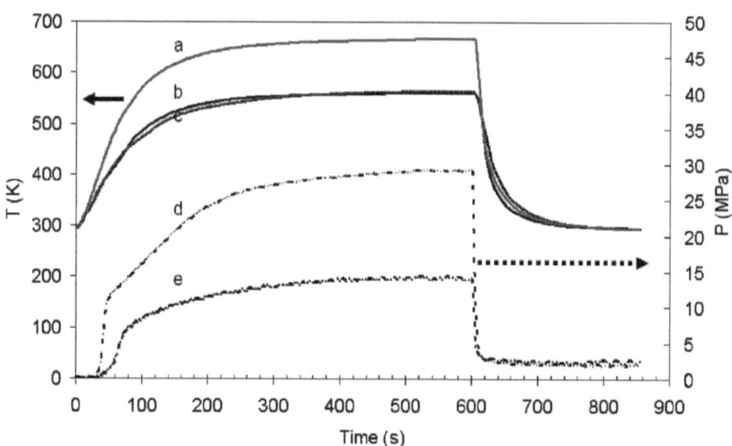

Fig. 2.8 Temperature and pressure profiles at different fluidized bed temperature (FBT) with or without H_2O_2 (49 wt%, 1 mL): **(a)** T at FBT = 673 K, **(b)** T at FBT = 573 K, **(c)** T at FBT = 573 K without H_2O_2, **(d)** P at FBT = 673 K, and **(e)** P at FBT = 573 K

Batch reactors are convenient for water-insoluble samples and long reaction time synthesis {e.g., several hours vs. 1 min in a flow reactor [24]} but not suitable for fast synthesis or some inorganic salt systems [25]. For example, the hydrolysis of acetate leads to the formation of acetic acid (boiling point of 391 K), which causes incontrollable increase of the pressure in the system. The temperature and concentration gradients as well as the slow rates for both heating and cooling lead to a non-uniform particle size distribution. But, the main advantages of batch reactors are (i) easy to design and operate; (ii) simple to increase reaction time for long time experiments {e.g., 25 h [26]}; and (iii) possible to control the oxidation states of the elements [25]. For example, copper hydroxy-carbonate decomposes into copper (II) oxide [25]:

$$(CuOH)_2CO_3 \rightarrow 2\ CuO\downarrow + CO_2 + H_2O \tag{2.8}$$

Oxalic acid under the same conditions decomposes with the formation of carbon monoxide:

$$H_2C_2O_4 \rightarrow CO + CO_2 + H_2O \tag{2.9}$$

The combination of these two reactions in SCW (658 K and 35 MPa) results in products of metallic copper (in the case of an excess of oxalic acid), a mixture of metallic copper and cuprous oxide (at salt/acid molar ratio $= 1/2$) and a mixture of copper (I) and copper (II) oxides (at salt/acid molar ratio $= 3/1$). The content of each phase is determined by the amount of oxalic acid added and, thus, can easily be controlled. Such a technique can be very efficient for preparation of catalysts and magnetic materials.

2.3 Flow Reactor

In contrast to batch systems, continuous flow reactors allow a better control of experimental conditions (e.g., temperature, pressure, time, concentration, pH and heating rate).

Figures 2.9 and 2.10 show a typical flow process and reactor used in Adschiri et al.'s work [16, 17] for hydrothermal crystallization of oxide nanoparticles. An aqueous metal salt solution [e.g., $Al(NO_3)_3$] is prepared and fed into the apparatus in one stream (Fig. 2.10: 4–6). In another stream, distilled water is pressurized and then heated to a temperature that is above the temperature desired (Fig. 2.10: 1–3, 6). The pressurized metal salt solution stream and the pure water stream are then combined in a mixing point (Fig. 2.10). This leads to rapid heating and subsequent reaction in the reactor (Fig. 2.10: 6). After the solution leaves the reactor, it is rapidly quenched by cold water and larger particles are removed by filters (Fig. 2.10: 6–11).

Pressure is controlled with a back-pressure regulator (Fig. 2.10: 10). The fine particles are collected in the effluent. Particles are separated and analyzed using the similar ways described in the above batch process. Conversion rate (%) is calculated

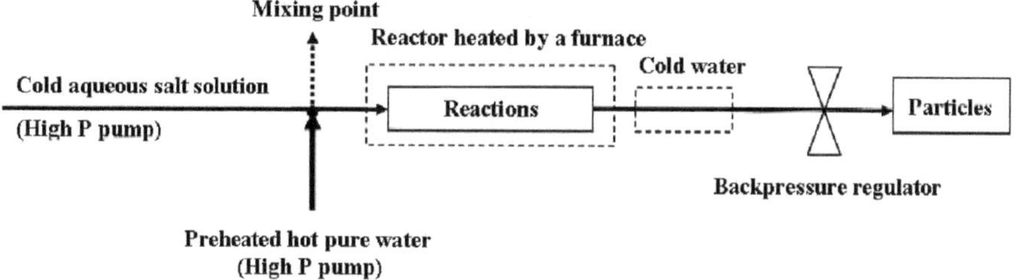

Fig. 2.9 A flow continuous process for rapid synthesis of fine particles in SCW

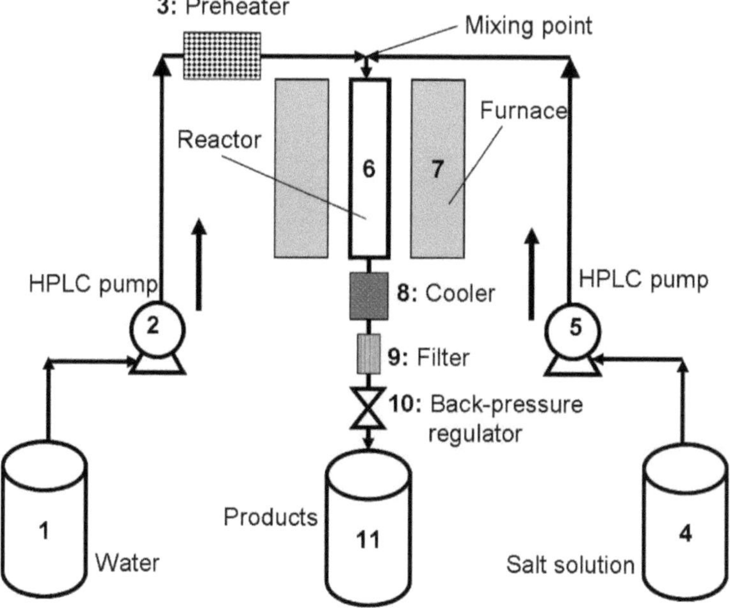

Fig. 2.10 A continuous flow reactor for rapid synthesis of fine particles in SCW

by dividing the difference between initial and final concentration of metal cation by its initial concentration. Productivity of the metal oxide particles is determined by the feed rate of the feed solution and conversion rate [21].

The reaction time (τ; min) is calculated by dividing the reactor volume (V; mL) by the flow rate at the reaction temperature (Q_r; mL/min):

$$\tau = V/Q_r \tag{2.10}$$

$$\tau = V(\rho_r/\rho_0)/Q_0 \tag{2.11}$$

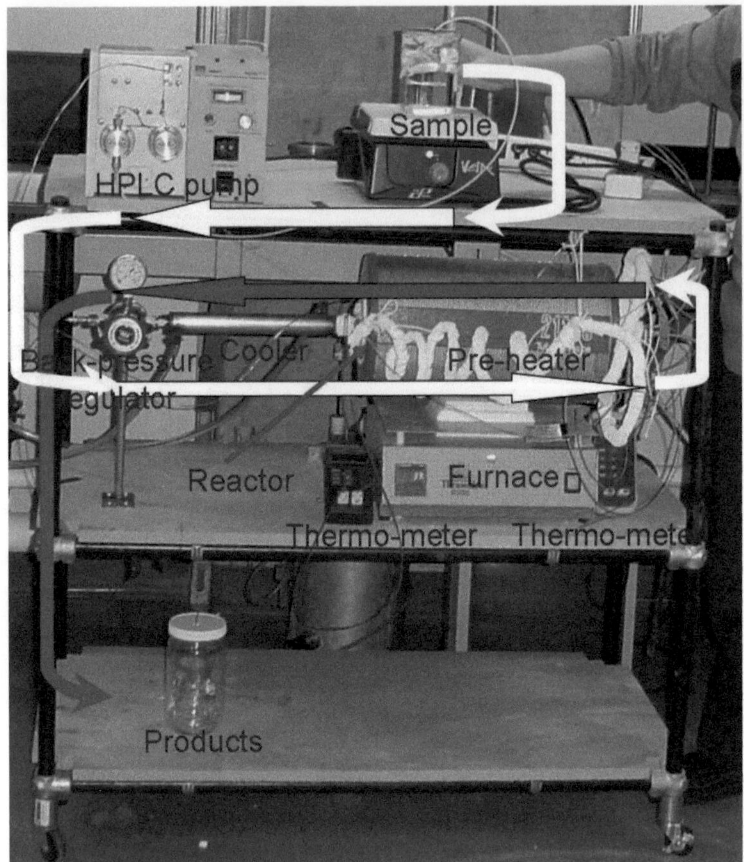

Fig. 2.11 Image of the flow reactor built in our lab

where ρ_r represents the solution density at reaction temperature (kg/m^3), ρ_0 is the initial solution density (kg/m^3), and Q_0 is the initial flow rate (mL/min). Usually, τ is less than 2 min.

Figure 2.11 shows the image of a continuous flow reactor built in our lab with only one sample stream, without the preheated pure water stream (Fig. 2.10: 1–3).

2.4 Diamond Anvil Cell (DAC)

Another type of reactor, an optical micro-reactor (50 nL), Bassett-type diamond anvil cell (DAC; Figs. 2.12, 2.13, 2.14, 2.15, and 2.16) [27] that is widely used in hydrothermal systems for the study of minerals, polymers and biomass can be easily applied to the in-situ visual and micro-spectroscopic study of particle production by

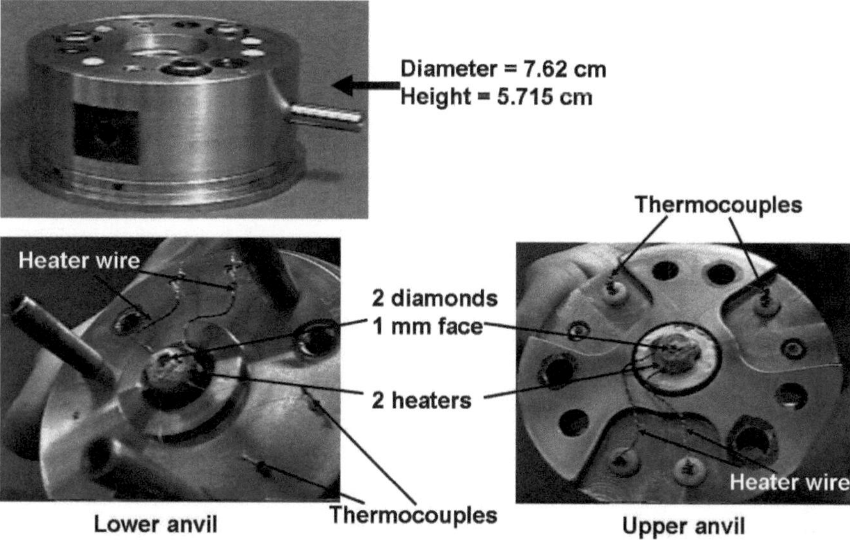

Fig. 2.12   Image of Bassett-type diamond anvil cell for the study of particle production in SCW

chemical synthesis and physical precipitation of solute from supercritical solution [27–30]. The DAC technique has the following merits: (i) clear visualization of the entire sample; (ii) access to spectroscopic techniques of FTIR/Raman/XRD to in-situ monitor reactions; (iii) high heating and cooling rates ($\pm$ 20 K/s) due to small

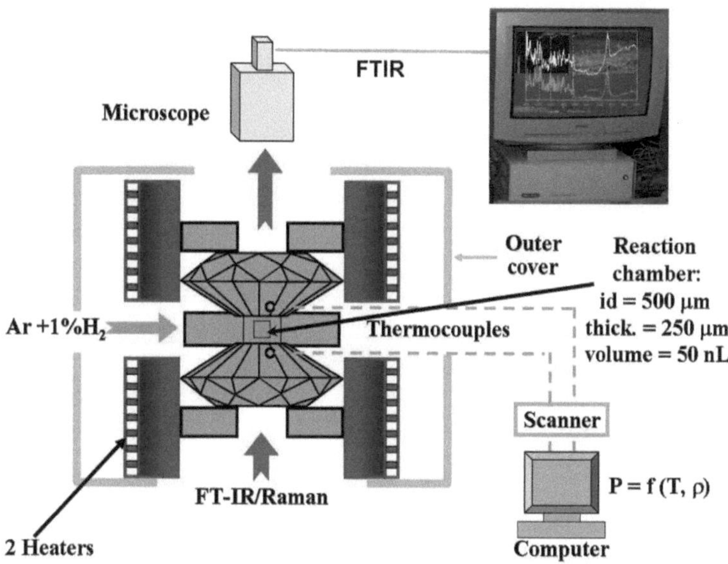

Fig. 2.13   Diamond anvil cell set-up system

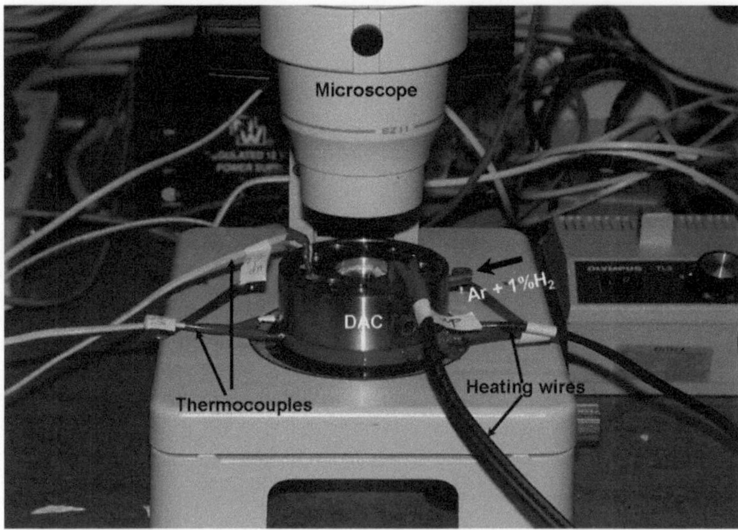

Fig. 2.14 Image of the DAC set-up ready for an experiment

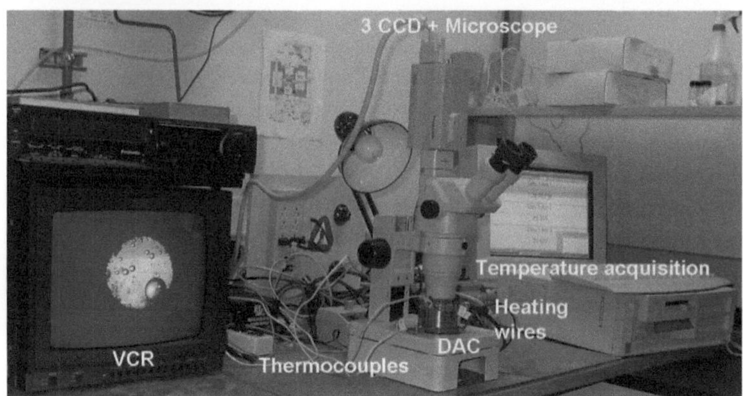

Fig. 2.15 Image of the DAC set-up system in experiment in our lab

volume; (iv) relatively inert materials in contact with the sample (diamonds, inconel gasket and ruby or ^{13}C diamond pressure sensor).

The DAC can be readily applied to hydrothermal systems at pressures up to 30 GPa and temperatures up to 1,473 K. Pressure is produced by two opposing diamond anvils inside the 50 nL chamber hole (internal diameter 508 μm, thickness 250 μm) made of inconel gasket (Fig. 2.13). Samples (solid) are loaded by placing solution (salt solutions, pure water, co-solvents) droplet into the gasket hole with a micro-syringe. The screw nuts of the DAC are tightened alternately in a 4-degree increment per turn, during which the samples are examined by an optical microscope (Olympus SZX16-3131) to check the loading. The samples and solution are heated

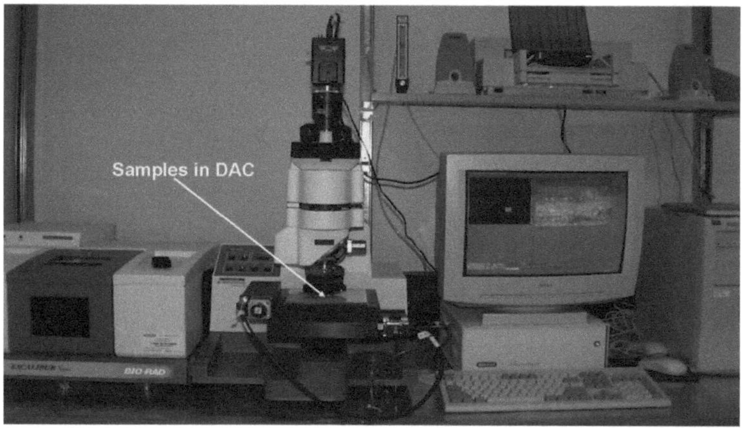

Fig. 2.16 FTIR micro-spectroscopy for analyzing samples from DAC

by two individual micro-heaters that transfer heat to the diamond anvils. Argon or nitrogen gas with 1% H_2 is introduced into the cell to protect the diamond anvils from oxidization and to increase the cooling rates at the end of the experiment.

Temperature is measured by two K-type thermocouples attached to each of the diamond anvils, and is recorded every 0.1 s by a data acquisition system (HP, Model 34970A). The thermocouples are calibrated by loading powder form samples contained in a gasket of reference metals tin ($T_m = 505.00$ K) and zinc ($T_m = 692.58$ K) and observing their melting points T_m with a slow heating rate (< 5 K/min).

The accuracy of the temperature measurement is estimated to be ± 0.5 K for temperatures up to 673 K, and ± 1.0 K for temperatures above 673 K. The temperature difference between the anvils is generally below 10 K (temperatures reported are the average of the both anvils) [31]. Pressure is determined from an EOSW [23] based on the negligible changes of chamber volume of the DAC, which was confirmed by Bassett [27] through the observation of the interference fringes between anvil faces radiated by a green light ($\lambda = 535$ nm). When the sample and water are loaded into the DAC chamber, Ar or N_2 gas bubbles will be introduced. Heating the chamber causes the liquid to expand and the gas bubbles to shrink until they disappear, at which point the chamber is filled with the expanded liquid at the homogenization temperature (T_h). The pressure at this point (P_h) is the vapor pressure along the liquid-vapor (L–V) curve of H_2O at T_h, and the bulk specific volume (υ_h) of the water is that of the liquid water along the L–V curve at T_h. If heated further, pressure will increase according to the P–T path of an isochore ($\upsilon = 1/\rho = \upsilon_h = $ chamber $-$ volume/water $-$ mass; m^3/kg). Thus, pressure can be calculated by knowing T at υ_h. The above calculation is assumed that only water contributes to the pressure. Figure 2.17 shows an example of temperature measured and pressure calculated in DAC.

At high water density ($> 1,000$ kg/m^3), pressure can be measured by ruby fluorescence R-lines of a ruby chip loading in the gasket hole [31, 32] with the equations

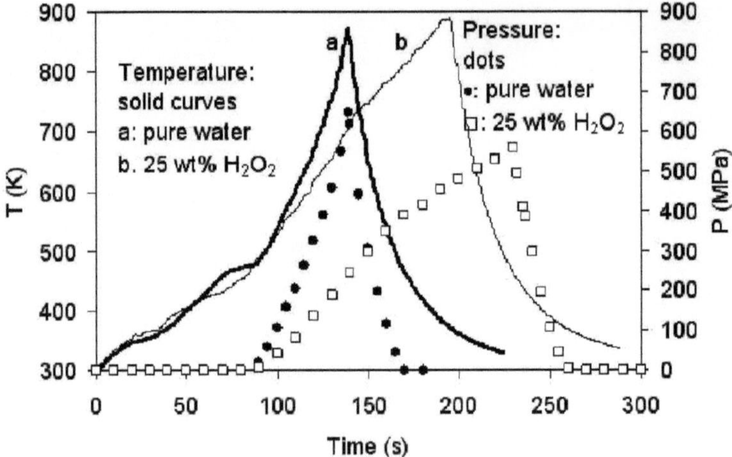

Fig. 2.17 Temperature (*solid curves*) measured and pressure (*dots*) calculated profiles vs. time in DAC with pure water (water density $= 867$ kg/m^3) and 25% H_2O_2 (water density $= 917$ kg/m^3)

of Forman et al. [33]:

$$P_1 = -(\varpi_1 - \varpi_{01})/0.77 \tag{2.12}$$

$$P_2 = -(\varpi_2 - \varpi_{02})/0.84 \tag{2.13}$$

where P and ϖ are the pressure and frequency of the ruby R_1 and R_2 lines having units of kbar and cm^{-1}, respectively. The $(\varpi_1 - \varpi_{01})$ and $(\varpi_2 - \varpi_{02})$ refer to the wave-number shift that occurs when the ruby is taken from room temperature and ambient pressure to the temperature and pressure of interest.

The ϖ is commonly determined by absolute methods, but relative values were used in our work [31, 32]. The fluorescence R-lines are conveniently measured with the Raman monochromator and the argon ion laser source. Pressure can be determined from either Eq. (2.12) or Eq. (2.13), but in our work, we used both R_1 and R_2 lines and determined pressure by averaging the values [31].

At higher temperatures, the formulas can be still used, provided the temperature dependent coefficient $\Delta\varpi/\Delta T$ is known. The coefficient has been shown to be independent pressure. In our experiments, we calibrated the coefficient over the range of 298~696 K by measuring the shift change with temperature at atmospheric pressure, and the value used was [32]:

$$\Delta\varpi/\Delta T = 0.1363 \ \text{cm}^{-1}/\text{K} \tag{2.14}$$

At temperatures higher than 818 K, the ruby R-lines disappear and ruby reacts slowly in SCW [34]. This reaction can be expected to increase at the higher temperatures. Therefore, pressure calibrated ^{13}C diamond can be used as the pressure sensor for these cases due to its inertness under high temperature conditions. The equation used to determine the pressure is from Schiferl et al. [34]:

$$P = (\varpi - \varpi_0)/2.83 \text{ GPa} \tag{2.15}$$

where ϖ_0 is the Raman wavenumber (cm^{-1}) of ^{13}C diamond at ambient pressure and room temperature, and ϖ is the corresponding frequency at the given pressure.

The DAC screw nuts are adjusted prior to experiments to obtain the initial pressure desired. Samples (e.g., salt solution, biomass) are heated rapidly (e.g., 10 K/s) or slowly (e.g., 0.2 K/s) and observed at 110× magnification, and the images are recorded by a Panasonic 3CCD camera (AW-E350) in a computer or video cassette recorder (Figs. 2.14 and 2.15), and are analyzed with digital imaging to calculate the apparent concentration (area %) and phase transition and crystal growth rates. Reaction can be monitored by putting DAC in a microscope coupled with FT-IR (Fig. 2.16) or Raman spectroscopy (Varian 610-IR/Renishaw 3000 Raman microscope). After heating for a pre-determined time, the sample is rapidly cooled by cutting power to the heaters while maintaining Ar gas flow. The residues deposited on the diamond faces are analyzed by FT-IR/Raman microscopy and SEM. Details of experimental setup and procedures can be seen in our previous works summarized in the review paper [28] and book [29].

References

1. T. Masui, Y. Peng, K. Machida, G. Adachi, Reduction behavior of CeO_2–ZrO_2 solid solution prepared from cerium zirconyl oxalate. Chem. Mater. **10**, 4005–4009 (1998)
2. M. Yashima, K. Morimoto, N. Ishizawa, M. Yoshimura, Zirconia-ceria solid solution synthesis and the temperature-time-transformation diagram for the 1:1 composition. J. Am. Ceram. Soc. **76**, 1745–1750 (1993)
3. C. de Leitenburg, A. Trovarelli, F. Zamar, S. Maschio, G. Dolcetti, J. Llorca, A novel and simple route to catalysts with a high oxygen storage capacity: The direct room-temperature synthesis of CeO_2–ZrO_2 solid solutions. J. Chem. Soc. Chem. Commun. **21**, 2181–2182 (1995)
4. A. Deptula, M. Carewska, T. Olczak, W. Lada, F. Croce, Sintering of zirconia-ceria spherical powders prepared by a water extraction variant of the sol-gel process. J. Electrochem. Soc. **140**, 2294–2297 (1993)
5. S. Hirano, T. Yogo, K. Kikuta, E. Asai, K. Sugiyama, H. Yamamoto, Preparation and phase separation behavior of cobalt iron oxide (($Co,Fe)_3O_4$) films. J. Am. Ceram. Soc. **76**, 1788–1792 (1993)
6. J.F. Hochepied, P. Bonville, M.P. Pileni, Nonstoichiometric zinc ferrite nanocrystals: syntheses and unusual magnetic properties. J. Phys. Chem. B **104**, 905–912 (2000)
7. T. Masui, K. Fujiwara, Y. Peng, T. Sakata, K. Machida, H. Mori, G. Adachi, Characterization and catalytic properties of CeO_2–ZrO_2 ultrafine particles prepared by the microemulsion method. J. Alloys Compd. **269**, 116–122 (1998)
8. M. Kiyama, The formation of manganese and cobalt ferrites by the air oxidation of aqueous suspensions and their properties. Bull. Chem. Soc. Jpn. **51**, 134–138 (1978)
9. T. Kodama, Y. Wada, T. Yamamoto, M. Tsuji, Y. Tamaura, Synthesis and characterization of ultrafine nickel(II)-bearing ferrites ($Ni_xFe_{3-x}O_4$, x = 0.14–1.0). J. Mater. Chem. **5**, 1413–1418 (1995)
10. B. Djuricic, D. McGarry, S. Pickering, The preparation of ultrafine ceria-stabilized zirconia particles coated with yttria. J. Mater. Sci. Lett. **12**(16), 1320–1323 (1993)

11. A. Cabanas, M. Poliakoff, The continuous hydrothermal synthesis of nano-particulate ferrites in near critical and supercritical water. J. Mater. Chem. **11**, 1408–1416 (2001)

12. G.W. Morey, P. Niggli, Hydrothermal formation of silicates, a review. J. Am. Chem. Soc. **35**, 1086–1130 (1913)

13. A. Rabenau, The role of hydrothermal synthesis in preparative chemistry. Angew. Chem. **97**, 1017–1032 (1985)

14. R. Gainsford, M.J. Sisley, T.W. Swaddle, P. Bayliss, Hydrothermal formation of ferrite spinels. Can. J. Chem. **53**, 12–19 (1975)

15. H. Kumazawa, K. Oki, H.M. Cho, E. Sada, Hydrothermal synthesis of ultrafine ferrite particles. Chem. Eng. Commun. **115**, 25–33 (1992)

16. T. Adschiri, K. Kanazawa, K. Arai, Rapid and continuous hydrothermal synthesis of boehmite particles in subcritical and supercritical water. J. Am. Ceram. Soc. **75**, 2615–2618 (1992)

17. T. Adschiri, K. Kanazawa, K. Arai, Rapid and continuous hydrothermal crystllization of metal oxides particles in supercritical water. J. Am. Ceram. Soc. **75**, 1019–1022 (1992)

18. K. Sue, N. Kakinuma, T. Adschiri, K. Arai, Continuous production of nickel fine particles by hydrogen reduction in near-critical water. Ind. Eng. Chem. Res. **43**(9), 2073–2078 (2004)

19. R.L. Smith Jr., P. Atmaji, Y. Hakuda, Y. Kawaguchi, T. Adschiri, K. Arai, Recovery of metals from simulated high-level liquid waste with hydrothermal crystallization. J. Supercrit. Fluids **11**(1,2), 103–114 (1997)

20. Z. Fang, S.K. Xu, J.A. Kozinski, Behavior of metals during combustion of industrial organic wastes in supercritical water. Ind. Eng. Chem. Res. **39**(12), 4536–4542 (2000)

21. T. Adschiri, Y. Hakuta, K. Arai, Hydrothermal synthesis of metal oxide fine particles at supercritical conditions. Ind. Eng. Chem. Res. **39**, 4901–4907 (2000)

22. Inorganic Chem. group, Dalian S&T university, *Inorganic Chemistry* (in Chinese) (Higher education press, Beijing, China, 1990), p. 645

23. W. Wagner, A. Pruss, The IAPWS formulation 1995 for the thermodynamic properties of ordinary water substance for general and scientific use. J. Phys. Chem. Ref. Data **31**(2), 387–535 (2002)

24. Y. Hakuta, T. Adschiri, T. Suzuki, T. Chida, K. Seino, K. Arai, Flow method for rapidly producing barium hexa-ferrite particles in supercritical water. J. Am. Ceram. Soc. **81**, 2461–2464 (1998)

25. A.A. Galkin, B.G. Kostyuk, N.N. Kuznetsova, A.O. Turakulova, V.V. Lunin, M. Polyakov, Unusual approaches to the preparation of heterogeneous catalysts and supports using water in subcritical and supercritical states. Kinetika i kataliz **42**, 172–181 (2001)

26. R.B. Yahya, H. Hayashi, T. Nagase, T. Ebina, Y. Onodera, N. Saitoh, Hydrothermal synthesis of potassium hexatitanates under subcritical and supercritical water conditions and its application in photocatalysis. Chem. Mater. **13**, 842–847 (2001)

27. W.A. Bassett, A.H. Shen, M. Bucknum, I.M. Chou, A new diamond-anvil cell for hydrothermal studies to 2.5 GPa and from −190 °C to 1200 °C. Rev. Sci. Instrum. **64**, 2340–2345 (1993)

28. R.L. Smith Jr., Z. Fang, Techniques, applications and future prospects of diamond anvil cells for studying supercritical water systems. J. Supercrit. Fluids **47**, 431–446 (2009)

29. Z. Fang, *Complete Dissolution and Oxidation of Organic Wastes in Water*, VDM Verlag Dr. Müller Aktiengesellschaft & Co. KG, Saarbrücken, Germany, ISBN 9783639144246, 192 pages, Apr. 2009.

30. H. Assaaoudi, Z. Fang, I.S. Butler, J.A. Kozinski, Synthesis of erbium hydroxide microflowers and nanostructures in subcritical water. Nanotechnolgy **19**, 185606 (8 pp) (2008)

31. Z. Fang, R.L. Smith Jr., H. Inomata, K. Arai, Phase behavior and reaction of polyethylene terephthalate-water systems at pressures up to 173 MPa and temperatures up to 490 °C. J. Supercrit. Fluids **15**, 229–243 (1999)

32. Z. Fang, R.L. Smith Jr., H. Inomata, K. Arai, Phase behavior and reaction of polyethylene in supercritical water at pressure up to 2.6 GPa and temperature up to 670°C. J. Supercrit. Fluids **16**, 207–216 (2000)

33. R.A. Forman, G.J. Piermarini, J.D. Barnett, S. Block, Pressure measurement made by the utilization of ruby sharp-line luminescence. Science **176**, 284–285 (1972)
34. D. Schiferl, M. Nicol, J.M. Zaug, S.K. Sharma, T.F. Cooney, S.-Y. Wang, T.R. Anthony, J.F. Fleischer, The diamond $^{13}C/^{12}C$ isotope Raman pressure sensor system for high-temperature/pressure diamond-anvil cells with reactive samples. J. Appl. Phys. **82** 3256–3265 (1997)

Chapter 3
Metal Oxides Synthesis

Abstract Many oxide/oxide composite nanocrystals were synthesized in continuous flow SCW processes in short time (e.g., 0.4 s~2 min) from metal cations via hydrolysis and subsequent dehydration. Metal particles could be produced by further reduction of the metal oxide in SCW. These particles included ferrite magnetic pigments [Fe_3O_4, MFe_2O_4 (M = Co, Ni, Zn), $Ni_xCo_{1-x}Fe_2O_4$, $BaO\cdot6Fe_2O_3$] for recording media, YAG: Tb phosphor for cathode ray tube screen, materials ($LiCoO_2$, $LiMn_2O_4$) for lithium ion battery cathode, catalysts for car exhausts [$Ce_{1-x}Zr_xO_2$ (x = 0~1), $Zr_{1-x}In_xO_2$, $Zr_{1-x}Y_xO_2$], oxidation (La_2CuO_4) and gasification (ZrO_2, CeO_2, Ni), photocatalytic materials ($K_2Ti_6O_{13}$, ZnO, TiO_2) for water decomposition, materials (SnO_2, ZnO, In_2O_3) for electronics industry, supporting materials (AlOOH) for catalysts, and many other oxides. Oxides were also produced in batch reactors for long reaction time but with larger size. Organic capping ligands were successfully used to control and stabilize particles size (TiO_2, Fe_2O_3, Cu, CeO_2). Oxide nanoparticles could also be synthesized via other chemical reactions in SCW.

3.1 Introduction

As early as 1992, Arai and Adschiri group for the first time synthesized metal oxide particles continuously in SCW in a flow reactor (Fig. 2.10) in less than 2 min [1–2]. In their work, ten metal aqueous salts [$Al(NO_3)_3$, $Fe(NO_3)_3$, $Fe_2(SO_4)_3$, $FeCl_2$, $Fe(NH_4)_2H(C_6H_5O_7)_2$, $Co(NO_3)_2$, $Ni(NO_3)_3$, $ZrOCl_2$, $TiSO_4$ and $TiCl_4$] with concentrations of 0.0066~0.16 mol/L (at sample feeding rate of 0.8~2.2 mL/min and water stream flow rate of 2.5~6 mL/min) were used as the starting materials to synthesize their corresponding metal oxides at conditions of 673~763 K and 30~35 MPa. Seven metal oxide nanocrystals (AlOOH: 600 nm; α-Fe_2O_3: 50 nm; Fe_3O_4: 50 nm; Co_3O_4: 100 nm; NiO: 200 nm; ZrO_2: 10 nm and TiO_2: 20 nm) were obtained with hexagonal, rhombic, needle, spherical, octahedral, rod-like and prismatic shapes. Later, they also produced other oxides [3–6]. In recent years, many works have been done on SCW production of oxides and other metal salts particles. Each individual oxide synthesis is discussed in details next.

Z. Fang, *Rapid Production of Micro- and Nano-particles Using Supercritical Water*, 29
Engineering Materials 30, DOI 10.1007/978-3-642-12987-2_3,
© Springer-Verlag Berlin Heidelberg 2010

3.2 Boehmite (AlOOH)

Boehmite (AlOOH) or pseudo-boehmite powders are used in the preparation of catalysts, coatings, alumina and alumina derived materials of desired porosity and mechanical strength [7]. In sub- and super-critical water, the effects of temperatures (523~763 K), pressures (25~40 MPa) and initial concentrations of $Al(NO_3)_3$ aqueous solution (0.0059~0.05 mol/L, feeding rate of 0.8~2.2 mL/min and water flow rate of 3.5~6 mL/min) on the particle size, morphology and crystal structure of boehmite were studied [1]. It was found that rhombic particles were formed at low temperatures and low concentrations while hexagonal crystals were obtained at high temperatures or concentrations. As temperature increased to 523 K, the shape became football-like with some laminar plates. Just above the critical temperature, needle-like particles were obtained. However, the effect of concentrations of solution on the particle size was not significant as temperatures and pressures. In the later work, a model was established based on the Gibbs energy change by temperatures, solvent effects and ion-ion interactions [8]. Using this model, the distribution of chemical species for boehmite system [Al^{3+}, $Al(OH)^{2+}$, $Al(OH)_2^{+}$, $Al(OH)_3$, $Al(OH)_4^{-}$, NO_3^{-}] in SubCW (623 K, 30 MPa) and SCW (673 K, 30 MPa) was predicted. The particle morphology seemed to be determined by selective adsorption of positive charged species, $Al(OH)^{2+}$, on the negatively charged faces of boehmite crystal. The mechanism of nanoparticle formation was discussed [9]. In the work of Hakuta et al. [10], 70~470 nm rhombic or hexagonal γ-AlO(OH) plates were produced as compared to larger particles (600 nm) obtained in the previous work [1]. The particle size increased with an increase in the reaction temperatures and the starting concentrations of $Al(NO_3)_3$ solution while decreased with an increase in the pH values of the starting solution. Moreover, at 673 K, the particle size enlarged from 170 to 300 nm with the increasing of reaction pressures from 25 to 40 MPa. Above 673 K, γ-Al_2O_3 nanoparticles (about 4 nm) were formed [11]. When $CH_3(CH_2)_4CHO$ and $CH_3(CH_2)_5NH_2$ were added in SCW, they were chemically bonded onto the surface of the AlOOH nanoparticles, which affected crystal growth and reduced the particle size thus changed the morphology of the particles [12]. For long reaction times (up to 92 h) in batch reactors, micron crystalline corundum was produced [13–14]. Increasing water pressures led to upsizing of corundum crystals and broadening of the crystal size distribution (20~80 μm at 21 MPa vs. 20~115 μm at 26 MPa). AlOOH (35~75 nm) and α-Al_2O_3 (75~130 nm) particles were also synthesized in SCW from alumina trihydrate according to the following reaction [15]:

$$2\ Al_2O_3 \cdot 3H_2O \rightarrow 2\ AlOOH + \alpha\text{-}Al_2O_3 + 2\ H_2O \qquad (3.1)$$

Reactive solution of $Al(NO_3)_3$ was investigated in the DAC (Fig. 2.12) by direct visual observation during rapid heating to SCW. A number of gelatinous phase transitions were observed that seem to play a key role in boehmite formation [16].

3.3 Ferrites

Pure magnetite (Fe_3O_4) and ferrites have been used in a variety of materials including magnetic pigments in recording media and magnetic materials for the storage and/or retrieval of information, as well as catalysts [17]. There is much interest in the production of nano-particulate magnetic materials, as reducing the particle size will allow the miniaturisation of devices. Besides the synthesis of 50-nm α-Fe_2O_3 and Fe_3O_4 particles by Arai and Adschiri group [3, 18], Cabanas and Poliakoff [19] obtained magnetic spinel type oxides such as magnetite (Fe_3O_4), ferrites of cobalt, nickel and zinc [MFe_2O_4 (M = Co, Ni, Zn)], and the mixed ferrite of nickel and cobalt [$Ni_xCo_{1-x}Fe_2O_4$] by the hydrolysis and simultaneous oxidation of the mixtures of Fe (II) acetate (Ac) and $M(II)(Ac)_2$ (0.1 M; Fe/M = 1/1, 2/1, 4/1 at flow rate of 5~10 mL/min for both salt and water streams) in sub- and super-critical water using a flow reactor. Possible reaction mechanisms were proposed as showed in Fig. 3.1.

Nilsen et al. [20] found the size of ferrites was between 39 and 105 nm with inverse spinel structures, which extended to the nano-regime. Sato et al. [21] obtained solid-solution nanoparticles of $MeFe_2O_4$ and γ-Fe_2O_3 with a cubic spinel structure and an average particle size under 10 nm from aqueous solutions of $Fe(NO_3)_3$ and $Me(NO_3)_2$ (Me = Ni, Cu, Zn) in SCW. In other work, α-Fe_2O_3 nanoparticles were synthesized and deposited successfully in activated carbon in SCW, which offers promise for carbon-supported catalyst preparation without the use of toxic or noxious solvents [22]. Synthesis of $CoFe_2O_4$ particles has been studied extensively in flow [23–24] and batch reactors [25], and 5-nm particles were produced at 663 K [26]. Millot et al. [26] used spark plasma sintering method to keep the powder properties in a bulk material.

Barium hexaferrite ($BaO\cdot6Fe_2O_3$) particles are hexagonal, plate-like, and have the magnetization axis perpendicular to the plate plane [27]. The perpendicular magnetizing recording method has been used to produce high-density recording

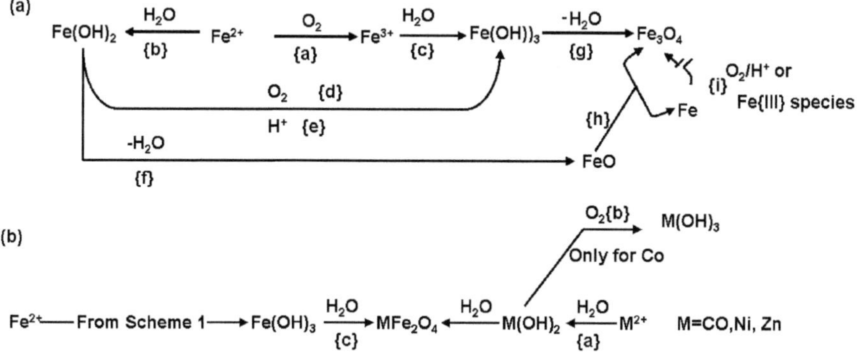

Fig. 3.1 Mechanism for the hydrothermal decomposition: (**a**) $Fe(Ac)_2$ and (**b**) a mixture of $Fe(Ac)_2$ and different metal(II) acetates, $M(Ac)_2$ [19]

media. Hydrothermal synthesis method is simpler than the current commercial glass-crystallization process [27–28]. In Adschiri group's work [28], barium hexaferrite particles were synthesized from aqueous iron (III) (0.02, 0.05, 0.1 M) and barium nitrate solutions [Ba/Fe = (0.1∼2)/1 at flow rate of 3 mL/min] in SCW using both a flow reactor (30 MPa and 673 K; less than 1 min) and a batch reactor (30 MPa and 653 K; 10∼240 min; Ba/Fe = 1). In the flow process, a flow rate at 3 mL/min for KOH solution and at 15 mL/min for water was used to adjust the solution's pH. When the Ba/Fe molar ratio was 1/12 (the stoichiometric ratio for $BaO·6Fe_2O_3$), the main product was α-Fe_2O_3. However, if Ba/Fe > 0.5/1, fine particles of single-phase $BaO·6Fe_2O_3$ were produced. The $BaO·6Fe_2O_3$ particles were hexagonal plates with size from 100 to 200 nm. Their magnetic properties are comparable with those of commercial products. Nam et al. [29] and Rho and Park [30] have obtained similar results in flow reactors. In the case of long time batch experiments, a mixture of α-Fe_2O_3 and $BaO·6Fe_2O_3$ was formed at conditions of Ba/Fe = 0.5/1 and 10-min reaction time. The mixture was changed to $BaO·6Fe_2O_3$ for longer reaction time (4 h). This suggests that the mixture of α-Fe_2O_3 and $BaO·6Fe_2O_3$ formed at short reaction time in the flow reactor was an intermediate phase [28]. In other work, smaller, uniform-size and single-phase $BaFe_{12}O_{19}$ nanocrystals were synthesized [31]. Drofenik et al. [32] obtained crystalline $BaFe_{12}O_{19}$ platelets with 50 nm in length and 5 nm in thickness that exhibited a saturation magnetization of 40 Am^2/kg.

3.4 Phosphor (YAG)

Aluminium yttrium garnet (YAG; $A_{l5}Y_3O_{12}$) based phosphors, such as YAG: Cr, YAG: Tb, YAG: Ce and YAG: Tm, are widely used in cathode ray tube. The commercial method for YAG phosphor synthesis is via SSR of stoichiometric mixtures of the component oxides at high temperature (2,273∼2,473 K), which needs long reaction times (e.g., 10∼20 h) [33]. Adschiri group has successfully produced a single phase of YAG: Tb particles at 673 K, 30 MPa and 1-min reaction time [33–34]. The particles were produced in a flow reactor by rapid heating stoichiometric mixtures of metal salt solutions [0.05−M $Al(NO_3)_3$, 0.03∼0.1 M $Y(NO_3)_3$ and $TbCl_3$ (Tb/Y = 1/20), at a flow rate of 1 mL/min] and an alkali solution {molar ratio of $[OH^-]/([NO_3^-]+[Cl^-]) = 1/1$, at a flow rate of 2 mL/min} with preheated hot water (at a flow rate of 8 mL/min). It was found that slow heating led to the formation of various intermediates (e.g., AlOOH, $AlYO_3$). However, rapid heating could prevent the formation of these intermediates. In other work, the particle size of YAG: Tb became finer as pH or potassium nitrate concentration of the starting metal salt solution increased. By removing the coexisting ions (NO_3^-, K^+) from the metal salt solution, single phase of YAG: Tb particles with 20-nm particle size were produced [35]. Adding ethanol promoted crystallization and reduced the operating temperatures. From the variables studied, the most important parameter seemed to be temperatures, and much smaller particles (e.g., 50 nm) have been obtained under supercritical conditions in comparison with the larger size (e.g., 150 nm) under subcritical conditions [36]. On the other hand, YAG: Eu nanoparticles were also synthesized [37–39], and uniform and spherical particles with 74 nm were produced in a continuous reactor [38]. However, in batch-type reactors, needle-like or elliptical-like particles were formed with stronger emission intensity [39]. Y_2O_3: Eu luminescent nanoparticles with sphere and rod-shaped morphology (20~40 nm) were prepared in SCW using aqueous yttrium chloride/europium chloride solution, methanol, NaOH aqueous solution, urea, I-octadecene and oleic acid followed by calcination at 1,073 and 1,273 K [40]. The nanoparticles with sphere morphology exhibited stronger luminescence intensity than rod-shaped particles.

3.5 $LiCoO_2/LiMn_2O_4$

$LiCoO_2$, $LiMn_2O_4$ and $Li_{4/3}Ti_{5/3}O_4$, as rechargeable lithium battery electrodes, have been commercially used. Rhombohedral $LiCoO_2$ single-crystal particles were successfully synthesized in a flow reactor in SCW at 673 K, 30 MPa and less than 1-min reaction time [41–43]. LiOH (0.4 mol/L, flow rate of 1.6 mL/min), $Co(NO_3)_2$ (0.02 mol/L, flow rate of 0.08 mL/min) and H_2O_2 (as an oxidant at flow rate of 0.08 mL/min) were the starting materials. The $LiCoO_2$ particles had more developed crystallinity with smaller size (500~1,000 nm) as compared with those produced by the most used method, SSR. They were good enough as a cathode material with rechargeability, but showed slightly lower discharge capacity due to probable crystal defects and contamination by amorphous phase. When $Mn(NO_3)_2$ solution was used as a starting material instead of $Co(NO_3)_2$ solution, $LiMn_2O_4$ crysal particles were produced [42, 44]. However, their discharge capacity was small due to a presence of large amount impurity phases. On the other hand, the reversibility of discharge and charge processes was extremely high as compared with an ordinary $LiMn_2O_4$. An addition of H_2O_2 to the starting solution was effective for preparation of a single phase of $LiMn_2O_4$, and both particle size and crystallinity increased by the heat-treatment at 1,073 K [45]. It was suggested that the reaction temperature should be relatively higher than the critical temperature of water for synthesizing particles with uniform size and shape [46].

3.6 $Ce_{1-x}Zr_xO_2$ (x = 0~1)

Ceria (CeO_2) and ceria-containing materials is of considerable interest for applications as diverse as catalysts in car exhausts, glass polishing, ultraviolet absorbers, luminous materials, solid oxide fuel cells and ceramics. Compared with CeO_2, $Ce_{1-x}Zr_xO_2$ shows improved ability to store and release oxygen (oxygen storage capacity) and greater thermal stability [47–49]. Adschiri group produced 20-nm octahedral ceria particles with 99.9% conversion rate from cerium nitrate (0.05 mol/L at a feeding rate of 6 mL/min, with water flow rate of 6 mL/min) at 673 K, 30 MPa and 0.4~20 s reaction time [49]. Recently, it was found that the addition of a heat-stable biopolymer (lignosulfate) resulted in the drastic downsizing of CeO_2 particle to 5 nm, because the lignosulfate functioned as an accelerator for the CeO_2 nanoparticle synthesis [50]. In other work, ceria nanoparticles (3~8 nm) were successfully produced and supported on multi-wall carbon nanotubes homogeneously without additional treatment [51]. Control of the morphology of ceria nanocrystals was achieved by tuning the reaction of organic ligand molecules [52].

The mechanism of nanoparticle formation at supercritical conditions was discussed based on the metal oxide solubility and kinetics of the hydrothermal synthesis reaction [9]. Cabanas et al. [48] synthesized nanocrystals (4~7 nm) of cubic CeO_2, {40% monoclinic + 60% tetragonal ZrO_2}, and cubic or tetragonal $Ce_{1-x}Zr_xO_2$ by hydrolysis of mixtures of cerium ammonium nitrate and zirconium acetate [0.21−M (Ce + Zr); Ce/Zr = 4/1, 1/1, 1/4 and 1/9] in near-critical water at 573 K and 25 MPa using a flow reactor. The composition was largely determined by the initial relative concentrations of Ce^{4+} and Zr^{4+} ions in the starting solution. The phases of $Ce_{1-x}Zr_xO_2$ (except Ce/Zr = 1/1) remained stable on calcining to 1,273 K, but the particles sintered and the surface areas decreased significantly. As compared with the conventional co-precipitation method, the supercritical synthesis could lead to ceria-zirconia mixed oxides with higher thermal stability and better oxygen storage capacity due to the morphology [53]. Zirconia nanocrystals (mean diameter of 6.8 nm) with tetragonal crystal structure could be formed from 0.05 M zirconyl acetate solution in the presence of 0.1 M potassium hydroxide at supercritical conditions [54]. Galkin et al. [47] also obtained 3~5 nm ZrO_2 with a 1/1 ratio mixture of tetragonal and monoclinic lattices at SCW conditions of 653 K, 25 MPa and 4.5-s reaction time. High crystal symmetry favors the mobility of oxygen in the volume of the ZrO_2 catalyst. A critical particle size of 5~6 nm for nanocrystalline monoclinic ZrO_2 was obtained in a flow reactor [55]. In order to stabilize the crystal lattices, some of transition and alkali-earth metals such as indium, yttrium and cerium were added, and $Zr_{1-x}In_xO_2$, $Zr_{1-x}Y_xO_2$, $Ce_{0.5}Zr_{0.5}O_2$, $Ce_{0.1}Y_xZr_{0.9-x}O_2$, 1% Rh-Pd/ZrO_2 and 10% Rh-Pd/ZrO_2 were synthesized at the same conditions. The catalytic activity of nanocrystalline Pd/ZrO_2 (TiO_2) samples prepared in sub- and super-critical water was studied for the oxidation of CO. They led to the complete conversion of CO at temperatures of 423~473 K that were much lower than those by using industrial Pd catalysts. The high activity of the nanocrystalline samples was caused by the presence of $Pd_xM_{1-x}O_2$ (M = Zr, Ti) solid solutions or intermetallic phases formed in the bulk ZrO_2 phase during synthesis [56]. In the Zr–In system, an increase of In resulted in the growth of the ratio of tetragonal to monoclinic modifications. However, the complete stabilization of the tetragonal structure was not achieved even In concentration increased up to 10%. The situation was improved much in the Zr–Y system. Adding of 10% Y concentration was sufficient for the predominant formation of the tetragonal modification of ZrO_2. Fluorescent nanoparticles of zirconia doped with europium (ZrO_2: Eu^{3+}) have been produced via the co-precipitation of simple metal salt precursors using continuous SCW hydrothermal synthesis [57]. Particles with small diameters (< 5 nm) at low temperatures but large diameters (> 20 nm) at high temperatures were produced from 473 to 673 K. Nanocrystalline $BaZrO_3$ was synthesized in SCW in a continuous way using barium precursors of $Ba(CH_3COO)_2$ and $Ba(NO_3)_2$ that led to the pure perovskite phase [58].

3.7 Potassium Hexatitanate, Potassium Niobate and Titania

Potassium hexatitanate (KTO; $K_2Ti_6O_{13}$ or $K_2O\cdot6TiO_2$) is relatively cheap fibrous material with thermal durability, chemical resistivity and dispersibility. Theses properties find their uses as reinforcing material for plastics, heat-insulating paints, and filter materials. KTO has tunnel structure, which finds applications in photocatalytic materials. Alkali-metal haxatitanate is one of the semiconducting photocatalysts, which has the capability of invoking a photocatalytic decomposition of water to produce H_2 and O_2 [59–61]. Using the tunnels to accommodate the active phase (e.g., Ru) can enhance the catalytic activity further [62].

Yahya et al. [59] synthesized KTO using potassium hydroxide and titanium tetraisopropoxide (molar ratio of K/Ti = 1/2) as starting materials in SubCW (623 K, 16 MPa) and SCW (673 K, 28 MPa; 723 K, 43 MPa) for 2, 5 and 25-h reaction times in an autoclave. The particles were compared with those synthesized by SSR method, that was carried out using potassium carbonate and titanium oxide (anatase) in a K/Ti molar ratio of 1.1/6 and heating at 1,403 K for 5 h. SEM and TEM observations revealed that the particulate morphology of products by hydrothermal synthesis is long and thin fibers (with diameter of 0.5 μm) while the particles by SSR method was mainly short thick fibers (with diameter of 3 μm) (Fig. 3.2; a vs. b).

Ru was loaded on KTO fibers by ion-exchange reaction, using a 0.1-wt% $RuCl_3$ aqueous solution, left overnight at room temperature in a shaker. The solid was filtered, dried at 323 K overnight, and calcined at 623 K for 2 h. All KTO samples were thermal stable up to 1,027 K by TGA and Differential Thermal Analysis (DTA). The surface areas then were analyzed and it was found that those from hydrothermal method were more than 10 times larger than those from SSR method (3.2 m^2/g).

Fig. 3.2 SEM micrographs of KTO as synthesized at 5-h reaction time: (**a**) supercritical water at 43 MPa and 723 K and (**b**) solid-state condition at 1,403 K. Reprinted with permission from [59]. Copyright © 2001, American Chemical Society

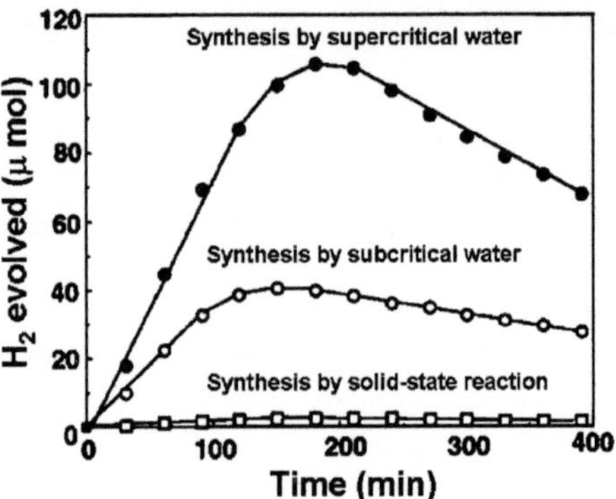

Fig. 3.3 Hydrogen evolution with time from water decomposition over various KTO/RuO$_2$ particles synthesized by: (o) subcritical water (673 K, 28 MPa, 5 h); (•) supercritical water (723 K, 44 MPa, 5 h); (□) solid-state reaction (1,403 K, 5 h). Reaction conditions: catalyst, 0.3 g; H$_2$O, 500 cm^3; a high-pressure mercury lamp (400 W); an inner irradiation type Pyrex cell. Reprinted with permission from [59]. Copyright © 2001, American Chemical Society

However, for the hydrothermal method, SCW conditions resulted in lower surface areas as compared to those produced in SubCW. The photo-catalytic activity of KTO with RuO$_2$ was tested for water decomposition reaction by a 400-W mercury lamp. It was found that photo-catalysts by hydrothermal method have much higher activities than those by the SSR method. More than 13 times for the photocatalysts synthesized under SubCW and 27~59 times for those synthesized under SCW conditions for water decomposition were obtained (Fig. 3.3). Surprisingly, the particles with smaller surface area from SCW had higher reactivity than those from SubCW. Later it was found that the particles from SCW had higher crystallinity, that was confirmed by XRD [59].

In short time (1.8 s) SCW synthesis experiments in a flow reactor using starting materials of titanium hydroxide sols and potassium hydroxide, KTO nanowires with 10 nm width and 500~1,000 nm length were produced at 663 K and 30 MPa (Fig. 3.4) [63]. The nanowires that were well dispersed and suspended in

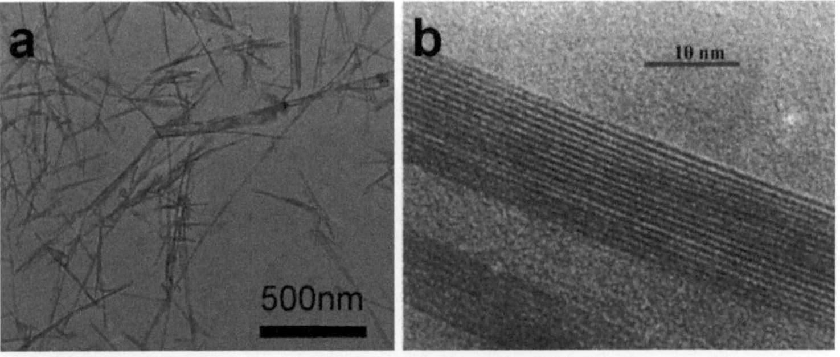

Fig. 3.4 TEM photographs of the KTO nanowires obtained at 673 K and 30 MPa in a flow reactor. Reprinted from [63]

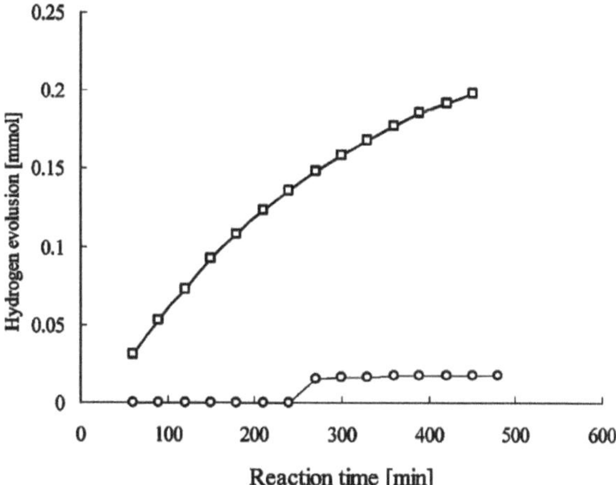

Fig. 3.5 Hydrogen evolution with time from methanol decomposition over KTO particles synthesized (□) in a flow reactor at 673 K and 30 MPa for 1.8 s, and (o) in batch reactor at 673 K and 28 MPa for 24 h. Reaction conditions: catalyst 0.3 g; 50 vol% methanol-H_2O, 500 cm^3, a high-pressure mercury lamp (400 W); an inner irradiation type Pyrex cell. Reprinted from [63]

water-methanol mixture were used in photo-decomposition of methanol for hydrogen production. The amount of hydrogen evolution was 10 times greater than those with the KTO particles that were produced by the above long time (24 h) SCW batch synthesis at 673 K and 28 MPa (Fig. 3.5) [63]. In other work, it is found that during the reforming of methanol with $KHCO_3$ in SCW, potassium titanate (K_2TiO_3) fibers were grown on the surface of a tubular reactor with a titanium liner that was apparently formed by reaction of Ti from the liner with $KHCO_3$ [64].

Potassium titanoniobate ($KTiNbO_5$) crystalline rectangular powders with large surface area for high photocatalytic performance were successfully synthesized [65–66]. The synthesized $KTiNbO_5$ powders with ruthenium loading were used for the photocatalytic hydrogen evolution from aqueous methanol solution [66]. A single phase of $K_4Nb_6O_{17}$ was formed under SubCW conditions, while mixed phases of $K_4Nb_6O_{17}$ and $KNbO_3$ were obtained under supercritical conditions [67]. The maximal hydrogen evolution rate was achieved for the potassium niobate synthesized at 673 K for 4 h. The rate was enhanced more than 10 times by Ni loading on potassium niobate powders, which was much higher in comparison with the Ni loaded solid-state synthesized photocatalyst. It was found that the $KNbO_3$ powders prepared in SCW with rhombohedral and orthorhombic structures exhibited strong intensity of second harmonic generation in the measuring of nonlinear optical properties [68].

Highly tetragonal crystalline $BaTiO_3$ particles around 50 nm in diameter were continuously produced at 673/693 K and 30 MPa within several seconds, whereas the particle size distribution was widely ranging from 10 to 150 nm. However, the finer $BaTiO_3$ particles at a size of 32 nm and a narrow size distribution could be produced by increasing the flow rate or decreasing reaction times [69]. When the density of water was less than 500 kg/m^3, the $BaTiO_3$ particles produced under the supercritical conditions had tetragonal phase [70]. As small as 10 nm $BaTiO_3$ particles were obtained at 673 K and 30 MPa [71]. In other work, ethanol and water were successfully used under supercritical conditions at 423∼653 K and 16 MPa in a continuous-flow reactor to synthesize ultra-fine barium titanate powders with a crystallinity as high as 90% and without barium carbonate contamination [72]. High pH value (pH > 13) is necessary to obtain phase-pure $BaTiO_3$ due to the solubility of titanium dioxide [73]. Rapid heating in a flow reactor is effective to synthesize smaller size and narrower particle size distribution as compared with the case of slow heating in a batch reactor.

Titania powders were synthesized in sub- and super-critical water using titanium (IV) tetraisopropoxide as a starting material. The average pore size of the titania powders increased from 4 to 25 nm with an increase in the crystallite size of the titania nanoparticles that aggregated to form the mesoporous structure. The titania powders were used for photocatalytic hydrogen evolution in an aqueous methanol suspension. Photocatalytic activity of the powders was found to be much higher in comparison with commercial titania photocatalysts [74]. When supercritical propanol-water was used, highly crystalline anatase TiO_2 nanoparticles were synthesized in less than 1 min in a continuous flow reactor [75]. The average particle size can be accurately controlled within 10~18 nm with narrow size distributions (2~3 nm). In comparison, a series of conventional autoclave sol-gel syntheses have been carried out. These also produced phase-pure anatase nanoparticles, but with much broader size distributions for much longer synthesis times.

3.8 Zinc Oxide

As an important material, zinc oxide particles find many applications, such as in varistors, semiconductors and photo-catalysts [76]. Spherical zinc oxide particles were obtained in the range of 39~320 nm in SCW in a continuous tubular reactor in less than 1 min from zinc acetate solution [76]. Nanosized zinc oxide particles were also produced by rapid heating zinc nitrate and potassium hydroxide aqueous solutions to supercritical conditions at 30 MPa and 573~673 K [77–78]. When formic acid solution was added, particles with hexagonal plates, hexagonal pillars and tube-like shapes were produced at formic acid concentration from 0.0 to 1.0 mol/kg. Relative intensity of emission spectra at around 500 nm increased from 0.53 to 1.0, the peak at 388 nm shifted to 393 nm, and the atomic ratio of O/(O + Zn) decreased from 0.49/1 to 0.38/1 [79]. Effects of cations and anions from $Zn(CH_3COO)_2$, $ZnSO_4$, $Zn(NO_3)_2$, LiOH and KOH on properties of zinc oxide particles were studied in a batch reactor [80]. It was found that particle size synthesized in LiOH solution was relatively smaller than that in KOH. Emission spectra of the particles produced from $ZnSO_4$ and LiOH aqueous solution showed the highest intensity among these systems. In a flow reactor at 659 K and 30 MPa, 16-nm particles were produced from $Zn(NO_3)_2$ and LiOH solution. Effect of inner diameter of a microreactor on particle size was examined at 673 K and 30 MPa. Nanocrystals with an average diameter of 9 nm were produced from {$ZnSO_4$ + KOH} solution using the micro-reactor with inner diameter of 0.3 mm [81]. Effects of heat-up order of {KOH + $Zn(NO_3)_2$} aqueous solution, concentrations, and flow rates on aspect ratio of ZnO crystals were studied, and nanorods with length of 230 nm and width of 38 nm were obtained [82]. Zinc oxide-titanium dioxide composite nanoparticles were prepared using an SCW reactor-in-series apparatus. Zinc oxide particles were produced from zinc acetate in the first reactor, and the particle suspension was fed to the second reactor where titanium oxide was formed from a titanium tetrachloride solution yielding ZnO–TiO_2 composite nanoparticles [83]. A new method was developed to synthesize zinc oxide nanoparticles via the reaction of solid zinc, $Zn(S)$ or liquid zinc, $Zn(L)$ with SCW and followed by their subsequent growth at $n > 7$ [84]:

$$Zn(S, L) + n\, H_2O = Zn(S, L) \cdot n(ZnO) + n\, H_2 \qquad (3.2)$$

$Zn(S)$ led predominantly to the formation of nanowires and nanorods, while the $Zn(L)$ practically always proceeded with the formation of nanoparticles. In other work, nanoparticles of zinc oxide-based materials (ZnO, $ZnAl_2O_4$) with various morphologies (rod, hexagonal, rectangular shapes) were synthesized in SCW. The synthesized nanomaterials were used as catalysts for the gasification of biomass to produce hydrogen [85].

3.9 Nickel, Nickel/Cobalt Oxide

Nickel fine particles with diameters under 1 μm are used in laminated ceramic capacitor and electrode materials [86]. Nickel and cobalt based nanomaterials are of interest in catalysis as inexpensive replacements for noble metal catalysts [87]. Nickel particles with diameters under 600 nm were synthesized rapidly and continuously in water at 573~653 K and 30 MPa. It is important to synthesize nanocatalysts that have magnetic properties to be easily separated from aqueous effluents after reaction for reuses in industry. Nickel particles coated on magnetic materials (e.g., Fe_3O_4) were formed through the following two steps: (i) formation of Fe_3O_4 nuclei by hydrothermal synthesis from $FeSO_4$ aqueous solution (Eqs. 2.1 and 2.2), and (ii) precipitation of nickel oxide on the surfaces of the nuclei (i.e., Ni coating) and subsequent hydrogen reduction to Ni by feeding both $Ni(CH_3COO)_2$ aqueous solution and hydrogen (Eqs. 2.1, 2.2, and 2.3) [86]. Because HCOOH can be used as a reductant, decomposing to CO and/or H_2 at higher temperatures, a new method for one-pot synthesis of Ni particles from nickel formate aqueous solution was proposed using an SCW environment at 673 K and 30 MPa in an autoclave (Eqs. 2.1, 2.2, and 2.3). High crystallinity and submicron size Ni particles were obtained [88]. Similarly, in a flow reactor, when HCOOH was used, smaller Ni nanoparticles (8.1 nm) were produced at 673 K in SCW by simultaneous reduction of NiO nanoparticles (3.8 nm), which were obtained from {$Ni(NO_3)_2$ + KOH} aqueous solution in the flow reactor at 673 K and 30 MPa [89].

A series of crystalline homometallic and heterometallic cobalt and nickel, hydroxides and oxides (typically less than 100 nm in diameter) were prepared using a continuous hydrothermal flow synthesis system [87]. Organic reagents such as R–COOH and R–NH$_2$ were used to in-situ modify organic surface of cobalt aluminate nanoparticles in SCW. The organic ligand capping could effectively inhibit the particle growth and also control the size of the nanocrystals [90].

3.10 La$_2$CuO$_4$

La$_2$CuO$_4$ and its derivatives are important catalysts used in the oxidation of CO [91–94] and CH$_4$ [92], decomposition of NO and N$_2$O [95–96] and NO reduction with CO [97–98]. Galkkin et al. [99, 47] synthesized La$_2$CuO$_4$ by two steps: (i) the formation of an intermediate mixture of La(OH)$_3$ and CuO from the decomposition of La(CH$_3$COO)$_3$ and Cu(CH$_3$COO)$_2$ in a continuous reactor at SCW conditions of 673~773 K and 25 MPa for a few seconds, and (ii) the single phase (La$_2$CuO$_4$)$_{sc}$ (subscript sc = SCW method) was obtained by sintering the mixture at 873 K for 5 h. For comparison, ceramic La$_2$CuO$_4$ was also synthesized by annealing a stoichiometric mixture of CuO and La$_2$O$_3$ at 1,373 K for 48 h with two intermediate grinding sessions. It was found that (La$_2$CuO$_4$)$_{sc}$ has much smaller and more uniform particle size with a very high BET surface area of up to 19 m^2/g as compared with only 0.4 m^2/g for ceramic La$_2$CuO$_4$. The catalytic tests showed that the activity of (La$_2$CuO$_4$)$_{sc}$ towards CO oxidation was 2.5 times higher than that of the ceramic La$_2$CuO$_4$ phase. Fine crystalline lanthanum aluminate was also formed in SCW [100]. In other work, NaLnGeO$_4$ (Ln = Ho, Tb, Tm, Er, Yb, Lu) was synthesized in SCW as a large single crystal, which belonged to the orthorhombic crystal system with the space group Pinna. The structures had olivine-type chains as the primary building block, with GeO$_4$ tetrahedra linking the chains through edge- and corner-sharing with lanthanide octahedral units [101]. Copper nanoparticles were continuously generated in SCW in a 316-SS tubular flow reactor at 973 K and 28 MPa from cupric acetate that was fed along with methanol by a HPLC pump [102]. Simultaneously, hydrogen was produced in-situ by reforming of methanol catalyzed by the generated nanoparticles in the reactor. The XRD pattern of the nanoparticles confirmed the presence of metallic copper along with cuprous and cupric oxides. Based on DLS analysis, the copper nanoparticles had a 140-nm mean diameter.

3.11 Organic-Inorganic Hybrid Nanocrystals

Ziegler et al. [103] used organic capping ligands to control and stabilize the synthesis of copper nanocrystals in SCW. In their work, in the presence of 1-hexanethiol [(CH$_3$(CH$_2$)$_5$SH; thiol/water mole ratio = 1/70] ligands, smaller copper nanocrystals (~7 nm in diameter, Fig. 3.6a) was produced from the hydrolysis of aqueous Cu(NO$_3$)$_2$ (0.02 M at 4 mL/min) in SCW (673 K and 41 MPa) in a flow reactor due to the particles being stabilized as compared with the formation of larger polydisperse copper(II) oxide particles (10~35 nm, Fig. 3.6b) without alkanethiol ligands. The use of a different precursor, Cu(CH$_3$COO)$_2$, led to the formation of

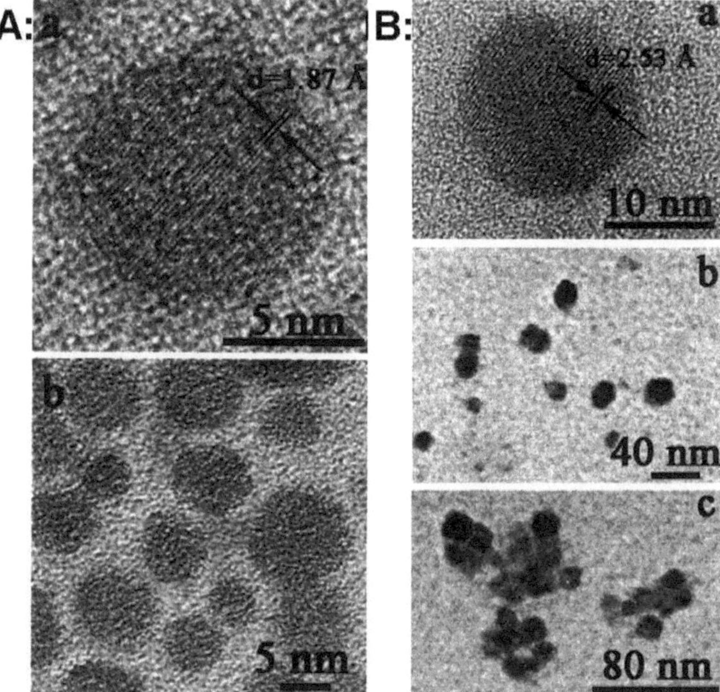

Fig. 3.6 TEM images of nanoparticles synthesized in SCW from $Cu(NO_3)_2$: **(a)** copper stabilized by 1-hexanethiol capping ligands and **(b)** CuO without capping ligands. Reprinted with permission from [103]. Copyright © 2001, American Chemical Society

significant different morphologic particles. Other processing conditions (e.g., pH, capping ligand) also greatly affected the morphology and size of nanoparticles.

The reaction mechanism was proposed for the capped copper particles in SCW (Fig. 3.7). Copper nitrate hydrolysis, in the absence of alkanethiol, may be described

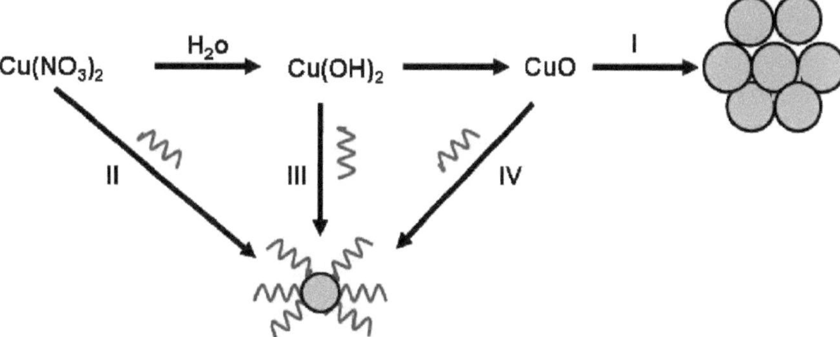

Fig. 3.7 The reaction mechanism for the formation of capped copper particles in SCW via $Cu(NO_3)_2$. Reprinted with permission from [103]. Copyright © 2001, American Chemical Society

by pathway I in Fig. 3.7, which led to nucleation of the oxide, CuO. In the presence of 1-hexanethiol, ligand exchange of thiol for anions and thiol-induced Cu(II) reduction could lead to Cu particles by pathways II, III, and possibly IV shown in Fig. 3.7. Pathways II and III competed with the hydrolysis/dehydration pathway I. CuO formed via pathway I could be reduced through the following mechanism (pathway IV):

$$2\,HSC_xH_{2x+1} + CuO \rightarrow (SC_xH_{2x+1})_2 + Cu + H_2O \tag{3.3}$$

$$HSC_xH_{2x+1} + Cu \rightarrow C_xH_{2x+1}SCu + 1/2\,H_2 \tag{3.4}$$

The alkanethiol, however, didn't appear to readily aid the reduction of Cu_2O to elemental copper when $Cu(CH_3COO)_2$ was used as the precursor for nanocrystal growth in SCW.

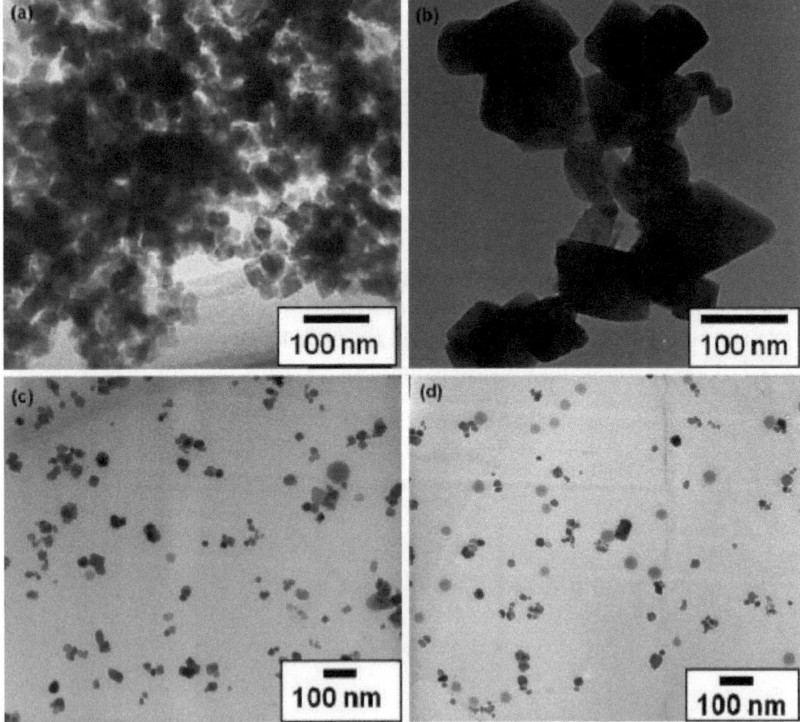

Fig. 3.8 TEM images of CuMn$_2$O$_4$ nanocrystals synthesized by (**a**) flow type reactor and (**b**) batch type reactor in the absence organic reagent; (**c**) batch type reactor in the presence of CH$_3$(CH$_2$)$_4$COOH and (**d**) in the presence of CH$_3$(CH$_2$)$_5$NH$_2$. Reprinted with permission from [106]. Copyright © 2008, Elsevier

Metal oxide nanoparticles (e.g., TiO_2, Fe_2O_3, Co_3O_4) whose surfaces were modified with organic materials (e.g., hexyl functional groups, decanoic acid) were successfully synthesized. In supercritical state, water and organic materials form a homogeneous phase, which provides an excellent reaction condition for the organic modification of nanoparticles. Modification with bio-materials including amino acids was also possible. By changing organic modifiers, particle morphology and crystal structure were changed [104]. A heat-stable biopolymer, lignosulfate, resulted in the drastic downsizing of CeO_2 particle to 5 nm [50]. Control of the morphology of ceria nanocrystals was achieved by tuning the reaction of organic ligand molecules (decanoic acid) with specific crystallographic planes of fluorite cubic ceria crystal during supercritical hydrothermal synthesis in a batch reactor at 673 K for 10 min [52, 105]. Changes in particle size and surface properties for other nanoparticles such as $CuMn_2O_4$ (Fig. 3.8) [106], $AlOOH$ [12] and $CoAl_2O_4$ [90] by surface modification were observed. It was found that adding polyvinyl alcohol (PVA) could limit the aggregation and reduce nanoparticle size of α-Fe_2O_3 [107] and $SrTiO_3$ [108] during hydrothermal synthesis. Photocatalytic activity of the PVA-assisted hydrothermally synthesized nanocrystalline $SrTiO_3$ powders for degradation of rhodamine B was much higher than those particles prepared without PVA [108].

3.12 Other Metal Oxides or Metals

Many other oxide fine particles such as V_2O_5 [109], $MgAl_2O_4$ [100], $CuMn_2O_4$ [106], strontium uranyl selenites {$Sr[(UO_2)_3(SeO_3)_2O_2]\cdot4H_2O$ and $Sr[UO_2(SeO_3)_2]$} [110], manganese oxide on alumina [111], silver on alumina [111–112], lead oxide on alumina [111], $Ba_{0.6}Sr_{0.4}TiO_3$ [113], partially substituted perovskite oxides {$Ca_{0.8}Sr_{0.2}Ti_{0.9}Fe_{0.1}O_{3-\alpha}$ [114] and $Ca_{0.8}Sr_{0.2}Ti_{1-x}Fe_xO_{3-\delta}$ [115] nanoparticles} and Mn-doped zinc silicate (Zn_2SiO_4: Mn^{2+}) [116–118] were synthesized in SCW in batch or flow reactors. Silver nanoparticles were produced by the thermal decomposition of Ag_2O in SCW in a batch reactor for long reaction times (up to 60 min) [119]. The size distribution of Ag nanoparticles was controllable in the range of 2~20 nm. Under appropriate conditions, the SCW medium

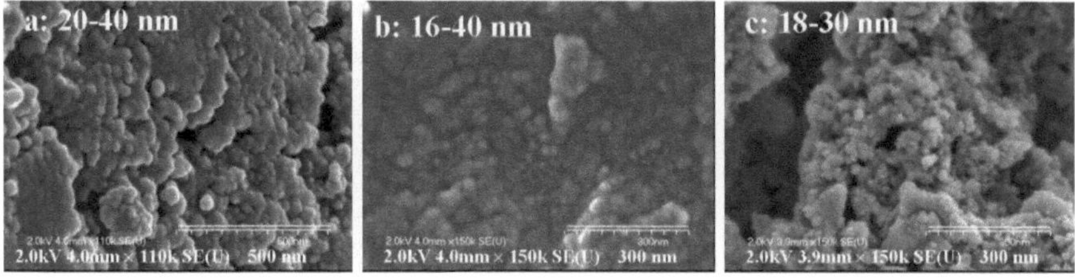

Fig. 3.9 SEM micrographs of the particles synthesized from aqueous solution of $SnCl_4$ in SCW and subsequent calcination at 723 K for 2 h, SCW conditions: (**a**) 663 K and 30 MPa for 53 s, (**b**) 688 K and 30 MPa for 51 s, and (**c**) 688 K and 30 MPa for 38 s

allowed the merging of Ag nanoparticles to form Ag nanowires with diameter of $\sim$60 nm, and nanobanners of triangle-shaped morphology with length of several hundred μm. In other work, well-dispersed Ru nanoparticles were successfully deposited on carbon nanotubes (CNT) with outer diameter of 40$\sim$60 nm and length of 1$\sim$12 μm from RuCl$_3$ in a batch reactor in SCW at 673 or 723 K for 2 h [120]. The general reaction can be expressed as:

$$4\ RuCl_3 + 3\ C + 6\ H_2O \rightarrow 4\ Ru\downarrow + 12\ HCl + 3\ CO_2 \tag{3.5}$$

Ru-CNT nanocomposites were subsequently demonstrated to be very active for the hydrogenation of benzene to cyclohexane with 100% conversion rate. Silver nanoparticles were also deposited on activated carbon [121].

Recently, we synthesized tetragonal SnO$_2$ and cubic In$_2$O$_3$ nanocrystals in SubCW and SCW from aqueous solutions of SnCl$_4$ and {SnCl$_4$ + InCl$_3$} in the flow reactor (Fig. 2.11) [122, 123]. They are considerable important materials in the electronics industry. The particles from aqueous solution of SnCl$_4$ seemed to have a large particle size (Fig. 3.9; 17$\sim$40 nm) by SEM due to agglomeration.

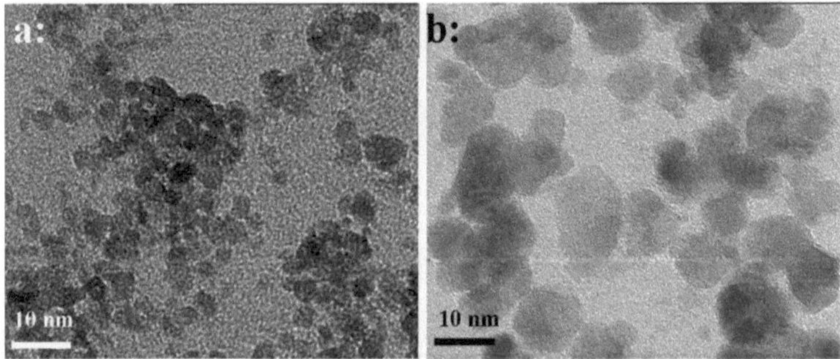

Fig. 3.10 TEM micrographs of the nanoparticles synthesized from aqueous solution of SnCl$_4$ in SCW at 663 K and 30 MPa for 53 s: (**a**) without calcination (3.7 nm; 3$\sim$5 nm) and (**b**) calcined at 873 K for 10 h (9 nm; 4$\sim$12 nm)

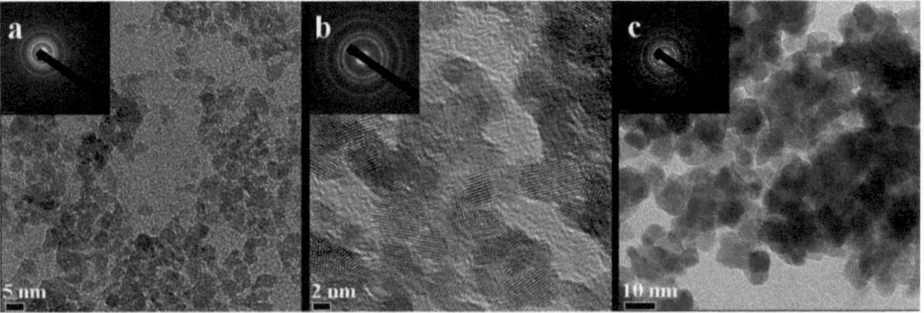

Fig. 3.11 Selected TEM micrographs and electron diffraction patterns of the nanoparticles synthesized from aqueous solution of SnCl$_4$ in SCW at 663 K and 30 MPa for 53 s. (**a**) and (**b**) without calcinations, and (**c**) calcined at 873 K for 10 h

However, after dispersed in alcohol solution by an ultrasonic machine, and subsequently analyzed by TEM, the actual size was about (3.7 nm) (Fig. 3.10a) and grew to 9 nm (Fig. 3.10b) after calcined at 873 K for 10 h. Some selected TEM micrographs and electron diffraction patterns of nanoparticles produced in SCW and SubCW were given in Figs. 3.11 and 3.12.

SEM-EDX elemental maps (Fig. 3.13) showed that the particles contained high level of Sn and O but lower concentration of Fe, Cl and Si. There was a close relationship between Sn and O, which showed that they were in the form of SnO_2. EDX spectra (Fig. 3.14) also confirmed existence of these elements but with an additional weak peak for Cr. Fe, Cr and Si were from the flow reactor that was made of 316-SS

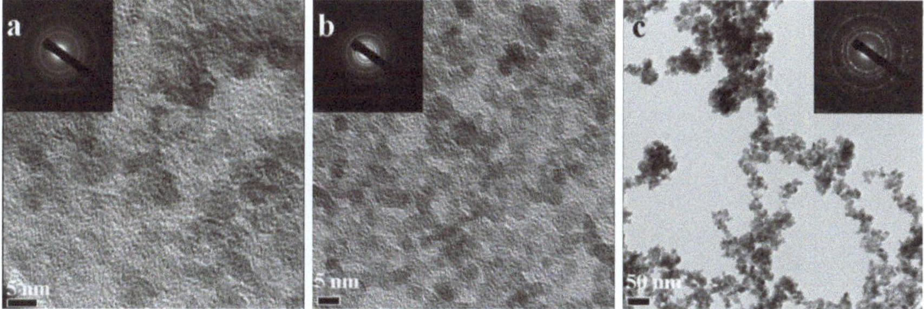

Fig. 3.12 Selected TEM micrographs and electron diffraction patterns of the nanoparticles synthesized from aqueous solution of $SnCl_4$ in SubCW at 623 K and 30 MPa for 109 s. (**a**) and (**b**) without calcination, and (**c**) calcined at 873 K for 10 h

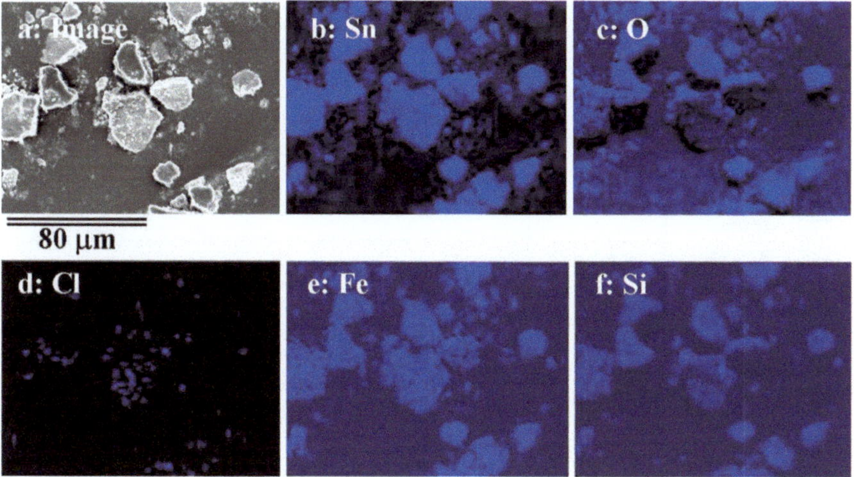

Fig. 3.13 EDX maps (Sn, O, Fe, Cl and Si) of the particles synthesized from aqueous solution of $SnCl_4$ in SCW (688 K and 30 MPa for 38.0 s) and calcined at 723 K for 2 h

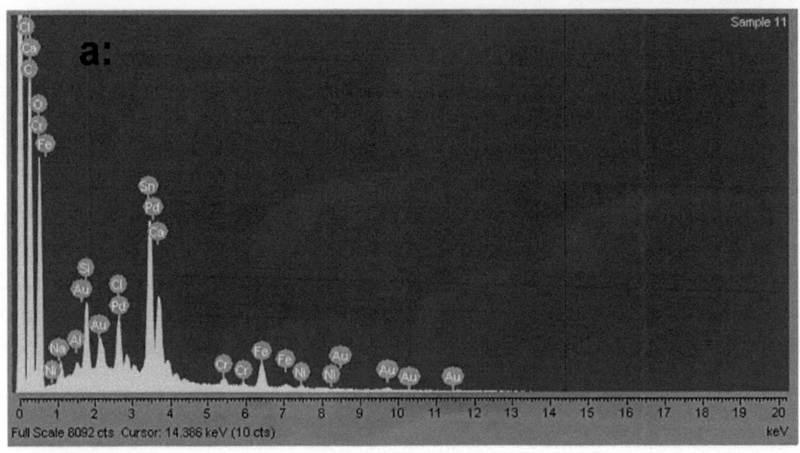
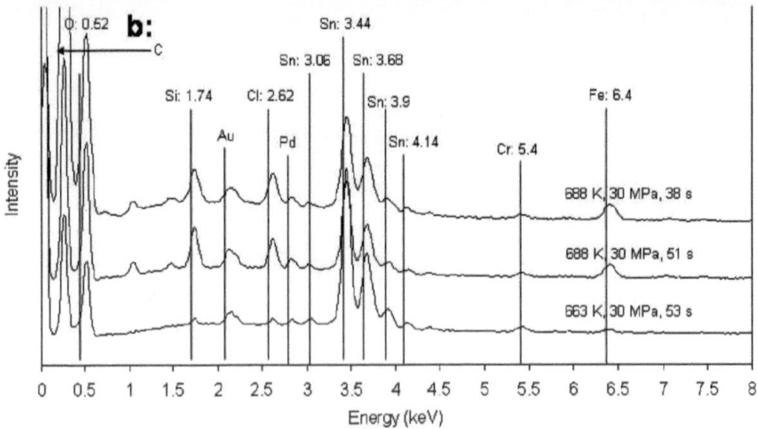

Fig. 3.14 EDX spectra of the synthesized particles from aqueous solution of SnCl₄ in SCW after calcined at 723 K for 2 h. (**a**) a selected sample and (**b**) the samples in Fig. 3.9a–c

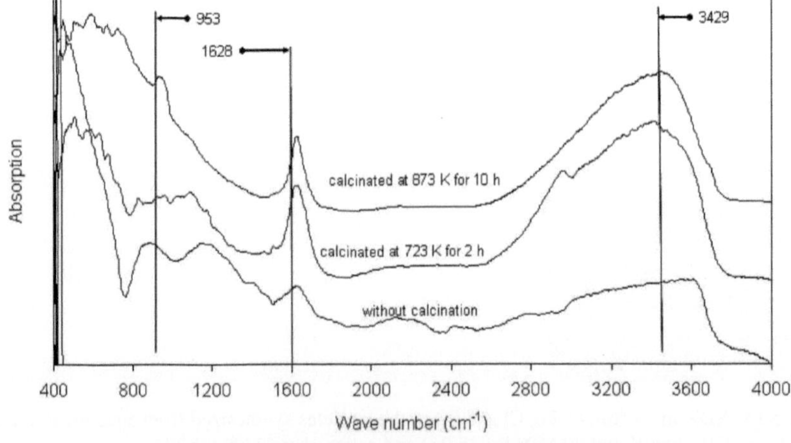

Fig. 3.15 IR spectra of the synthesized particles from aqueous solution of SnCl₄ in SCW (688 K and 30 MPa for 38.0 s) with and without calcination

tube due to corrosion. Cl was from the contamination from the initial $SnCl_4$ solution. Particles were also analyzed by FTIR and Raman as showed in Figs. 3.15 and 3.16.

XRD spectra showed that the particles were tetragonal SnO_2 nanocrystals (Fig. 3.17). The crystals are important in the electronics industry and can be used as a solid state gas sensor and a transparent battery electrode.

Nanocrystalline tin and indium oxides (In_2O_3/SnO_2) were synthesized in sub- and super-critical water at 623/653 K and 30 MPa in less than 73 s in a tubular

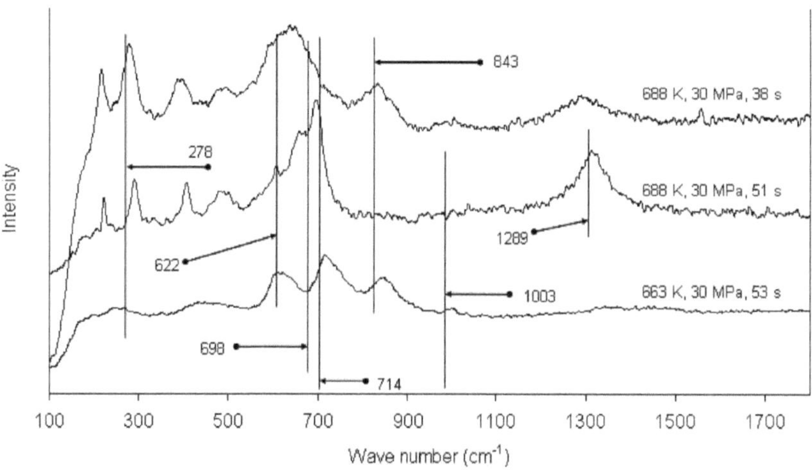

Fig. 3.16 Raman spectra of the synthesized particles from aqueous solution of $SnCl_4$ in SCW with calcined at 873 K for 10 h

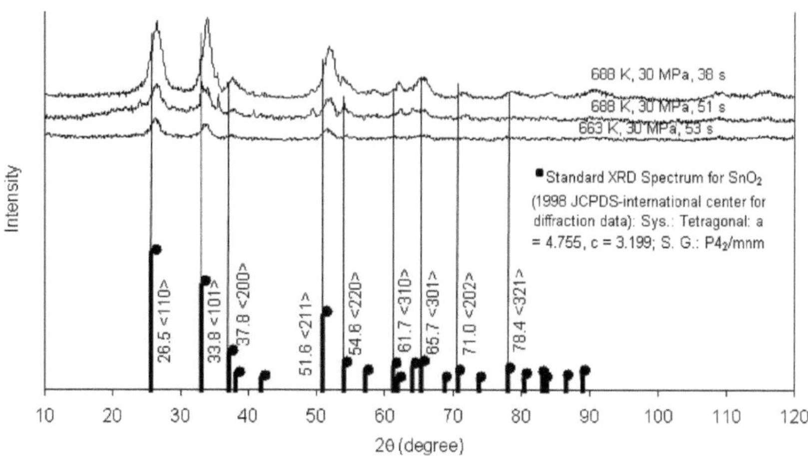

Fig. 3.17 XRD spectra of the synthesized particles from aqueous solution of $SnCl_4$ in SCW after calcined at 873 K for 10 h

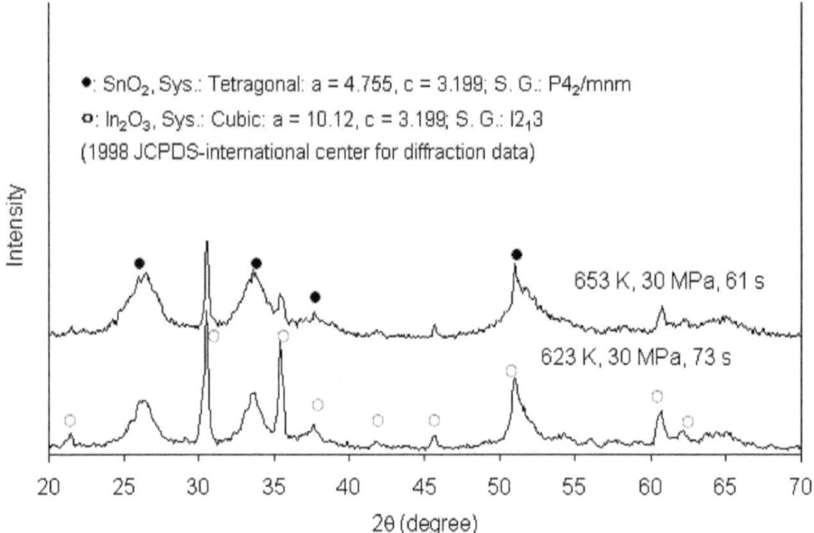

Fig. 3.18 XRD spectra of the synthesized nanoparticles from aqueous solution {SnCl$_4$ + InCl$_3$} in sub- and super-critical water

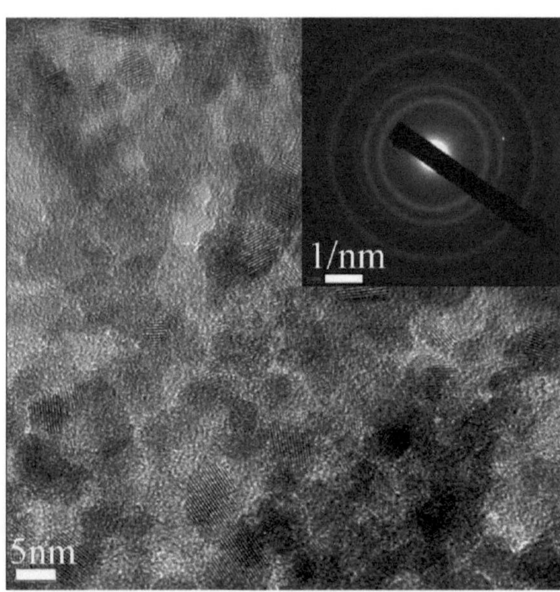

Fig. 3.19 TEM micrograph and electron diffraction pattern for the particles produced from aqueous solution {SnCl$_4$ + InCl$_3$} in SCW at 653 K, 30 MPa and 61 s. d (distance between two crystallographic plans) = 0.333, 0.267, 0.238, 0.177, 0.145 nm

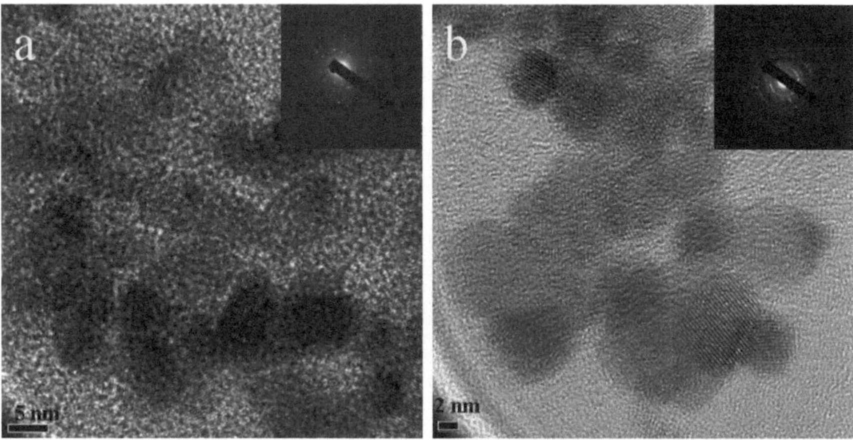

Fig. 3.20 Selected TEM micrographs and electron diffraction patterns of the nanoparticles synthesized from aqueous solution of {SnCl$_4$ + InCl$_3$} in (**a**) SCW at 653 K and 30 MPa for 61 s, and (**b**) SubCW at 623 K and 30 MPa for 73 s

flow reactor from an aqueous solution of {SnCl$_4$ + InCl$_3$} (Figs. 3.18, 3.19, and 3.20) [123]. The bulk particles were composed of In, Sn, and O atoms, and made up of cubic In$_2$O$_3$ (10 nm) and tetragonal SnO$_2$ (5.5 nm) crystals (Figs. 3.18, 3.19, and 3.20). From electron diffraction pattern in Fig. 3.19, it was found that minor (In$_{1-x}$Sn$_x$)OOH crystal was also formed, and would form tin-doped indium oxide (In$_2$Sn$_{1-x}$ O$_{5-y}$) particles.

References

1. T. Adschiri, K. Kanazawa, K. Arai, Rapid and continuous hydrothermal synthesis of boehmite particles in subcritical and supercritical water. J. Am. Ceram. Soc. **75**, 2615–2618 (1992)
2. T. Adschiri, K. Kanazawa, K. Arai, Rapid and continuous hydrothermal crystllization of metal oxides particles in supercritical water. J. Am. Ceram. Soc. **75**, 1019–1022 (1992)
3. T. Adschiri, Y. Hakuta, K. Arai, Hydrothermal synthesis of metal oxide fine particles at supercritical conditions. Ind. Eng. Chem. Res. **39**, 4901–4907 (2000)
4. Y. Hakuda, T. Ajiri, K. Arai, Hydrothermal synthesis of metal oxide fine particles using supercritical water. Chorinkai Saishin Gijutsu **4**, 11–15 (2000)
5. Y. Hakuta, T. Ajiri, K. Arai, Synthesis of micro-particle metal oxides by supercritical hydrothermal method. Kemikaru Enjiniyaringu **45**, 621–626 (2000)
6. T. Adschiri, Production of metal oxide fine particles via hydrothermal synthesis under supercritical conditions. Seramikkusu **35**, 534–537 (2000)
7. D. Mishra, S. Anand, R.K. Panda, R.P. Das, Effect of anions during hydrothermal preparation of boehmites. Mater. Lett. **53**, 133–137 (2002)
8. Y. Hakuta, T. Adschiri, H. Hirakoso, K. Arai, Chemical equilibria and particle morphology of boehmite (AlOOH) in sub and supercritical water. Fluid Phase Equilib. **158–160**, 733–742 (1999)
9. T. Adschiri, Y. Hakuta, K. Sue, K. Arai, Hydrothermal synthesis of metal oxide nanoparticles at supercritical conditions. J. Nanopart. Res. **3**(2–3), 227–235 (2001)

10. Y. Hakuta, H. Ura, H. Hayashi, K. Arai, Effects of hydrothermal synthetic conditions on the particle size of γ-AlO(OH) in sub and supercritical water using a flow reaction system. Mater. Chem. Phys. **93**(2–3), 466–472 (2005)

11. T. Noguchi, K. Matsui, N.M. Islam, Y. Hakuta, H. Hayashi, Rapid synthesis of γ-Al_2O_3 nanoparticles in supercritical water by continuous hydrothermal flow reaction system. J. Supercrit. Fluids **46**(2), 129–136 (2008)

12. T. Mousavand, S. Ohara, M. Umetsu, J. Zhang, S. Takami, T. Naka, T. Adschiri, Hydrothermal synthesis and in situ surface modification of boehmite nanoparticles in supercritical water. J. Supercrit. Fluids **40**(3), 397–401(2007)

13. M.N. Danchevskaya, S.N. Torbin, Y.D. Ivakin, G.P. Muravieva, Role of precursor compaction and water vapour pressure during synthesis of corundum in supercritical water. J. Phys. Condens. Matter. **16**, S1187–S1196 (2004)

14. M.N. Danchevskaya, Y.D. Ivakin, S.N. Torbin, G.P. Panasyuk, V.N. Belan, I.L. Voroshilov, Scientific basis of technology of fine-crystalline quartz and corundum. High Pressure Res. **20**(1–6), 229–239 (2001)

15. P.K. Panda, V.A. Jaleel, S.U. Devi, Hydrothermal synthesis of boehmite and alpha-alumina from Bayer's alumina trihydrate. J. Mater. Sci. **41**(24), 8386–8389 (2006)

16. T. Wang, R.L. Smith Jr., H. Inomata, K. Arai, Reactive phase behavior of aluminum nitrate in high temperature and supercritical water. Hydrometallurgy **65**(2–3), 159–175 (2002)

17. T. Kodama, Y. Wada, T. Yamamoto, M. Tsuji, Y. Tamaura, Synthesis and characterization of ultrafine nickel(II)-bearing ferrites ($Ni_xFe_{3-x}O_4$, x = 0.14–1.0). J. Mater. Chem. **5**, 1413–1418 (1995)

18. T. Adschiri, K. Kanazawa, K. Arai, Rapid and continuous hydrothermal crystllization of metal oxides particles in supercritical water. J. Am. Ceram. Soc. **75**, 1019–1022 (1992)

19. A. Cabanas, M. Poliakoff, The continuous hydrothermal synthesis of nano-particulate ferrites in near critical and supercritical water. J. Mater. Chem. **11**, 1408–1416 (2001)

20. M.H. Nilsen, C. Nordhei, A.L. Ramstad, D.G. Nicholson, XAS (XANES and EXAFS) investigations of nanoparticulate ferrites synthesized continuously in near critical and supercritical water. J. Phys. Chem. C **111**(17), 6252–6262 (2007)

21. T. Sato et al., Rapid and continuous production of ferrite nanoparticles by hydrothermal synthesis at 673 K and 30 MPa. Ind. Eng. Chem. Res. **47**(6), 1855–1860 (2008)

22. C.B. Xu, A.S. Teja, Supercritical water synthesis and deposition of iron oxide (α-Fe_2O_3) nanoparticles in activated carbon. J. Supercrit. Fluids **39**(1), 135–141 (2006)

23. L.J. Cote, A.S. Teja, A.P. Wilkinson, Z.J. Zhang, Continuous hydrothermal synthesis of $CoFe_2O_4$ nanoparticles. Fluid Phase Equilib. **210**(2), 307–317 (2003)

24. N. Millot, B. Xin, C. Pighini, D. Aymes, Hydrothermal synthesis of nanostructured inorganic powders by a continuous process under supercritical conditions. J. Eur. Ceram. Soc. **25**(12), 2013–2016 (2005)

25. D. Zhao, X. Wu, H. Guan, E. Han, Study on supercritical hydrothermal synthesis of $CoFe_2O_4$ nanoparticles. J. Supercrit. Fluids **42**(2), 226–233 (2007)

26. N. Millot, S. Le Gallet, D. Aymes, F. Bernard, Y. Grin, Spark plasma sintering of cobalt ferrite nanopowders prepared by coprecipitation and hydrothermal synthesis. J. Eur. Ceram. Soc. **27**(2–3), 921–926 (2007)

27. B.T. Shirk, W.R. Buessen, Magnetic properties of barium ferrite formed by crystallization of a glass. J. Am. Ceram. Soc. **53**, 192–196 (1970)

28. Y. Hakuta, T. Adschiri, T. Suzuki, T. Chida, K. Seino, K. Arai, Flow method for rapidly producing barium hexa-ferrite particles in supercritical water. J. Am. Ceram. Soc. **81**, 2461–2464 (1998)

29. S.C. Nam, S.S. Park, G.J. Kim, Preparation of Ba-ferrite particles using the supercritical water crystallization method. J. Ind. Eng. Chem. **7**, 38–43 (2001)

30. S. Rho, S. Park, Influence of stoichiometry and alkalinity on barium hexaferrite formation via the supercritical water crystallization method. Korean J. Chem. Eng. **19**, 120–125 (2002)

31. S.C. Nam, G.J. Kim, Characterization of barium hexaferrite produced by varying the reaction parameters at the mixing-points in a supercritical water crystallization process. Korean J. Chem. Eng. **21**(3), 582–588 (2004)
32. M. Drofenik, M. Kristl, A. Znidarsic, D. Hanzel, D. Lisjak, Hydrothermal synthesis of Ba-hexaferrite nanoparticles. J. Am. Ceram. Soc. **90**(7), 2057–2061 (2007)
33. Y. Hakuta, K. Seino, H. Ura, T. Adschiri, H. Takizawab, K. Arai, Production of phosphor (YAG: Tb) fine particles by hydrothermal synthesis in supercritical water. J. Mater. Chem. **9**, 2671–2674 (1999)
34. T. Haganuma, Y. Hakuta, T. Adschiri, K. Arai, Synthesis of phosphor (YAG: Tb) fine particles by hydrothermal synthesis on supercritical water. Koatsuryoku no Kagaku to Gijutsu **10**, 98 (2000)
35. Y. Hakuta, T. Haganuma, K. Sue, T. Adschiri, K. Arai, Continuous production of phosphor YAG: Tb nanoparticles by hydrothermal synthesis in supercritical water. Mater. Res. Bull. **38**(7), 1257–1265 (2003)
36. A. Cabanas, J. Li, P. Blood, T. Chudoba, W. Lojkowski, M. Poliakoff, E. Lester, Synthesis of nanoparticulate yttrium aluminum garnet in supercritical water-ethanol mixtures. J. Supercrit. Fluids **40**(2), 284–292 (2007)
37. M.J. Yoon, J.H. In, H.C. Lee, C.H. Lee, Comparison of YAG: Eu phosphor synthesized by supercritical water and solid-state methods in a batch reactor. Korean J. Chem. Eng. **23**(5), 842–846 (2006)
38. J.H. In, H.C. Lee, M.J. Yoon, K.K. Lee, J.W. Lee, C.H. Lee, Synthesis of nano-sized YAG: Eu^{3+} phosphor in continuous supercritical water system. J. Supercrit. Fluids **40**(3), 389–396 (2007)
39. M.J. Yoon, Y.S. Bae, S.H. Son, J.W. Lee, C.H. Lee, Comparison of YAG: Eu phosphors synthesized by supercritical water in batch and continuous reactors. Korean J. Chem. Eng. **24**(5), 877–880 (2007)
40. M.K. Devaraju, S. Yin, T. Sato, Solvothermal synthesis, controlled morphology and optical properties of Y_2O_3:Eu^{3+} nanocrystals. J. Cryst. Growth **311**(3), 580–584 (2009)
41. K. Kanamura, A. Goto, R.Y. Ho, T. Umegaki, K. Toyoshima, K. Okada, Y. Hakuta, T. Adschiri, K. Arai, Preparation and electrochemical characterization of $LiCoO_2$ particles prepared by supercritical water synthesis. Electrochem. Solid-State Lett. **3**, 256–258 (2000)
42. K. Kanamura, T. Umegaki, K. Toyoshima, K. Okada, Y. Hakuta, T. Adschiri, K. Arai, Electrochemical characteristics of $LiCoO_2$ and $LiMn_2O_4$ prepared in supercritical water. Key Eng. Mater. **181–182**, 147–150 (2000)
43. T. Adschiri, Y. Hakuta, K. Kanamura, K. Arai, Continuous production of $LiCoO_2$ fine crystals for lithium batteries by hydrothermal synthesis under supercritical condition. High Pressure Res. **20**(1–6), 373–384 (2001)
44. J.H. Lee, K.S. Jun, H.J. Choi, W.I. Cho, B.W. Cho, Synthesis of lithium manganese oxide particles in supercritical water. Kongop Hwahak **12**, 859–863 (2001)
45. K. Kanamura, K. Dokko, T. Kaizawa, Synthesis of spinel $LiMn_2O_4$ by a hydrothermal process in supercritical water with heat-treatment. J. Electrochem. Soc. **152**(2), A391–A395 (2005)
46. J.H. Lee, J.Y. Ham, Synthesis of manganese oxide particles in supercritical water. Korean J. Chem. Eng. **23**(5), 714–719 (2006)
47. A.A. Galkin, B.G. Kostyuk, N.N. Kuznetsova, A.O. Turakulova, V.V. Lunin, M. Polyakov, Unusual approaches to the preparation of heterogeneous catalysts and supports using water in subcritical and supercritical states. Kinetika i kataliz **42**, 172–181 (2001)
48. A. Cabanas, J.A. Darr, E. Lester, M. Poliakoff, Continuous hydrothermal synthesis of inorganic materials in a near critical water flow reactor; the one-step synthesis of nano-particulate $Ce_{1-x}Zr_xO_2$ (x = 0–1) solid solutions. J. Mater. Chem. **11**, 561–568 (2001)
49. Y. Hakuta, S. Onai, H. Terayama, T. Adschiri, K. Arai, Production of ultra-fine ceria particles by hydrothermal synthesis under supercritical conditions. J. Mater. Sci. Lett. **17**, 1211–1213 (1998)

50. M. Umetsu, X. Man, K. Okuda, M. Tahereh, S. Ohara, J. Zhang, S. Takami, T. Adschiri, Biomass-assisted hydrothermal synthesis of ceria nanoparticle – A new application of lignin as a bio-nanopool. Chem. Lett. **35**(7), 732–733 (2006)
51. D. Zhao, E. Han, X. Wu, H. Guan, Hydrothermal synthesis of ceria nanoparticles supported on carbon nanotubes in supercritical water. Mater. Lett. **60**(29–30), 3544–3547 (2006)
52. J. Zhang, S. Ohara, M. Umetsu, T. Naka, Y. Hatakeyama, T. Adschiri, Colloidal ceria nanocrystals: A tailor-made crystal morphology in supercritical water. Adv. Mater. **19**(2), 203–206 (2007)
53. J.R. Kim, W.J. Myeong, S.K. Ihm, Characteristics in oxygen storage capacity of ceria-zirconia mixed oxides prepared by continuous hydrothermal synthesis in supercritical water. Appl. Catal. B Environ. **71**(1–2), 57–63 (2007)
54. Y. Hakuta, T. Ohashi, H. Hayashi, K. Arai, Hydrothermal synthesis of zirconia nanocrystals in supercritical water. J. Mater. Res. **19**(8), 2230–2234 (2004)
55. J. Becker, P. Hald, M. Bremholm, J.S. Pedersen, J. Chevallier, S.B. Iversen, B.B. Iversen, Critical size of crystalline ZrO_2 nanoparticles synthesized in near- and supercritical water and supercritical isopropyl alcohol. ACS Nano **2**(5), 1058–1068 (2008)
56. A.A. Galkin, A.O. Turakulova, V.V. Lunin, J. Grams, The oxidation of CO and adsorption of hydrogen on nanocrystalline catalysts of the composition $Pd/ZrO_2(TiO_2)$ prepared in sub- and supercritical water. Russ. J. Phys. Chem. **79**(6), 881–885 (2005)
57. H. Hobbs, S. Briddon, E. Lester, The synthesis and fluorescent properties of nanoparticulate ZrO_2 doped with Eu using continuous hydrothermal synthesis. Green Chem. 11(4), 484–491 (2009)
58. A. Aimable, B. Xin, N. Millot, D. Aymes, Continuous hydrothermal synthesis of nanometric $BaZrO_3$ in supercritical water. J. Solid State Chem. **181**(1), 183–189 (2008)
59. R.B. Yahya, H. Hayashi, T. Nagase, T. Ebina, Y. Onodera, N. Saitoh, Hydrothermal synthesis of potassium hexatitanates under subcritical and supercritical water conditions and its application in photocatalysis. Chem. Mater. **13**, 842–847 (2001)
60. Y. Inoue, T. Kubokawa, K. Sato, Photocatalytic activity of alkali-metal titanates combined with ruthenium in the decomposition of water. J. Phys. Chem. **95**, 4059–4063 (1991)
61. T. Takata, Y. Furumi, K. Shinohara, A. Tanaka, M. Hara, J.N. Kondo, K. Domen, Photocatalytic decomposition of water on spontaneously hydrated layered perovskites. Chem. Mater. **9**, 1063–1064 (1997)
62. Y. Inoue, M. Kohno, K. Sato, S. Ogura, Photocatalytic activity for water decomposition of RuO_2-combined $M_2Ti_6O_{13}$ (M = Na, K, Rb, Cs). Appl. Surf. Sci. **121–122**, 521–524 (1997)
63. Y. Hakuta, H. Hayashi, K. Arai, Hydrothermal synthesis of photocatalyst potassium hexatitanate nanowires under supercritical conditions. J. Mater. Sci. **39**(15), 4977–4980 (2004)
64. W. Habicht, N. Boukis, G. Franz, O. Walter, E. Dinjus, Exploring hydrothermally grown potassium titanate fibers by STEM-in-SEM/EDX and XRD. Microsc. Microanal. **12**(4), 322–326 (2006)
65. B. Li, Y. Hakuta, H. Hayashi, Hydrothermal synthesis of crystalline rectangular titanoniobate particles. Chem. Commun. (13), 1732–1734 (2005)
66. B. Li, Y. Hakuta, H. Hayashi, Synthesis of potassium titanoniobate in supercritical and sub-critical water and investigations on its photocatalytic performance. J. Supercrit. Fluids **39**(1), 63–69 (2006)
67. H. Hayashi, Y. Hakuta, Y. Kurata, Hydrothermal synthesis of potassium niobate photocatalysts under subcritical and supercritical water conditions. J. Mater. Chem. **14**(13), 2046–2051 (2004)
68. B. Li, Y. Hakuta, H. Hayashi, Hydrothermal synthesis of $KNbO_3$ powders in supercritical water and its nonlinear optical properties. J. Supercrit. Fluids **35**(3), 254–259, (2005)
69. Y. Hakuta, H. Ura, H. Hayashi, K. Arai, Continuous production of $BaTiO_3$ nanoparticles by hydrothermal synthesis. Ind. Eng. Chem. Res. **44**(4), 840–846 (2005)

70. Y. Hakuta, H. Ura, H. Hayashi, K. Arai, Effect of water density on polymorph of $BaTiO_3$ nanoparticles synthesized under sub and supercritical water conditions. Mater. Lett. **59**(11), 1387–1390 (2005)
71. K. Matsui, T. Noguchi, N.M. Islam, Y. Hakuta, H. Hayashi, Rapid synthesis of $BaTiO_3$ nanoparticles in supercritical water by continuous hydrothermal flow reaction system. J. Cryst. Growth **310**(10), 2584–2589 (2008)
72. H. Reveron, C. Aymonier, A. Loppinet-Serani, C. Elissalde, M. Maglione, F. Cansell, Single-step synthesis of well-crystallized and pure barium titanate nanoparticles in supercritical fluids. Nanotechnology **16**(8), 1137–1143 (2005)
73. M. Atashfaraz, M. Shariaty-Niassar, S. Ohara, K. Minami, M. Umetsu, T. Naka, T. Adschiri, Effect of titanium dioxide solubility on the formation of $BaTiO_3$ nanoparticles in supercritical water. Fluid Phase Equilib. **257**(2), 233–237 (2007)
74. H. Hayashi, K. Torii, Hydrothermal synthesis of titania photocatalyst under subcritical and supercritical water conditions. J. Mater. Chem. **12**(12), 3671–3676 (2002)
75. P. Hald, J. Becker, M. Bremholm, J.S. Pedersen, J. Chevallier, S.B. Iversen, B.B. Iversen, Supercritical propanol-water synthesis and comprehensive size characterisation of highly crystalline anatase TiO_2 nanoparticles. J. Solid State Chem. **179**(8), 2674–2680 (2006)
76. R. Viswanathan, R.B. Gupta, Formation of zinc oxide nanoparticles in supercritical water. J. Supercrit. Fluids **27**(2), 187–193 (2003)
77. K. Sue, K. Murata, K. Kimura, K. Arai, Continuous synthesis of zinc oxide nanoparticles in supercritical water. Green Chem. **5**(5), 659–662 (2003)
78. S. Ohara, T. Mousavand, T. Sasaki, M. Umetsu, T. Naka, T. Adschiri, Continuous production of fine zinc oxide nanorods by hydrothermal synthesis in supercritical water. J. Mater. Sci. **43**(7), 2393–2296 (2008)
79. K. Sue, K. Kimura, K. Murata, K. Arai, Hydrothermal synthesis of zinc oxide crystals in homogeneous mixture of carbon dioxide, hydrogen, and water. Chem. Lett. **33**(6), 708–709 (2004)
80. K. Sue, K. Kimura, K. Murata, K. Arai, Effect of cations and anions on properties of zinc oxide particles synthesized in supercritical water. J. Supercrit. Fluids **30**(3), 325–331 (2004)
81. K. Sue, K. Kimura, K. Arai, Hydrothermal synthesis of ZnO nanocrystals using microreactor. Mater. Lett. **58**(25), 3229–3231 (2004)
82. K.W. Sue, K. Kimura, M. Yamamoto, K. Arai, Rapid hydrothermal synthesis of ZnO nanorods without organics. Mater. Lett. **58**(26), 3350–3352 (2004)
83. R. Viswanathan, G.D. Lilly, W.F. Gale, R.B. Gupta, Formation of zinc oxide-titanium dioxide composite nanoparticles in supercritical water. Ind. Eng. Chem. Res. **42**(22), 5535–5540 (2003)
84. A.A. Vostrikov, A.V. Shishkin, N.I. Timoshenko, Synthesis of zinc oxide nanostructures during zinc oxidation by sub- and supercritical water. Tech. Phys. Lett. **33**(1), 30–34 (2007)
85. A. Levy, M. Watanabe, Y. Aizawa, H. Inomata, K. Sue, Synthesis of nanophased metal oxides in supercritical water: Catalysts for biomass conversion. Int. J. Appl. Ceram. Technol. **3**(5), 337–344 (2006)
86. K. Sue, N. Kakinuma, T. Adschiri, K. Arai, Continuous production of nickel fine particles by hydrogen reduction in near-critical water. Ind. Eng. Chem. Res. **43**(9), 2073–2078 (2004)
87. P. Boldrin, A.K. Hebb, A.A. Chaudhry, L. Otley, B. Thiebaut, P. Bishop, J.A. Darr, Direct synthesis of nanosized $NiCo_2O_4$ spinel and related compounds via continuous hydrothermal synthesis methods. Ind. Eng. Chem. Res. **46**(14), 4830–4838 (2007)
88. K. Sue, A. Suzuki, M. Suzuki, K. Arai, Y. Hakuta, H. Hayashi, T. Hiaki, One-pot synthesis of nickel particles in supercritical water. Ind. Eng. Chem. Res. **45**(2), 623–626 (2006)
89. K. Sue, A. Suzuki, M. Suzuki, K. Arai, T. Ohashi, K. Matsui, Y. Hakuta, H. Hayashi, T. Hiaki, Synthesis of Ni nanoparticles by reduction of NiO prepared with a flow-through supercritical water method. Chem. Lett. **35**(8), 960–961 (2006)
90. D. Rangappa, S. Ohara, T. Naka, A. Kondo, M. Ishii, T. Adschiri, Synthesis and organic modification of $CoAl_2O_4$ nanocrystals under supercritical water conditions. J. Mater. Chem. **17**(41), 4426–4429 (2007)

91. S. Rajadurai, J.J. Carberry, B. Li, C.B. Alcock, Catalytic oxidation of carbon monoxide over superconducting lanthanum strontium copper oxide ($La_{2-x}Sr_xCuO_{4-\delta}$) systems between 373–523 K. J. Catal. **131**, 582–589 (1991)

92. R. Doshi, C.B. Alcock, N. Gunasekaran, J.J. Carberry, Carbon monoxide and methane oxidation properties of oxide solid solution catalysts. J. Catal. **140**(2), 557–563 (1993)

93. N. Gunasekaran, A. Meenakshisundaram, V. Srinivasan, Catalytic oxidation of carbon monoxide on potassium tetrafluoronickelate(II)-type perovskites – Lanthanum copper oxide (La_2CuO_4) and lanthanum nickel oxide (La_2NiO_4). Indian J. Chem. Sect. A **21A**, 346–349, (1982)

94. H. Yasuda, Y. Fujiwara, N. Mizuno, M. Misono, Oxidation of carbon monoxide on $LaMn_{1-x}Cu_xO_3$ perovskite-type mixed oxides. J. Chem. Soc. Faraday Trans. **90**, 1183–1189 (1994)

95. S. Subramanian, C.S. Swamy, Catalytic decomposition of nitrous oxide on strontium-substituted La_2CuO_4 materials. Catal. Lett. **35**, 361–372 (1995)

96. H. Yasuda, T. Nitadori, N. Mizuno, M. Misono, Catalytic decomposition of nitrogen monoxide over valency-controlled La_2CuO_4-based mixed oxides. Bull. Chem. Soc. Jpn. **66**, 3492–3502 (1993)

97. C. Oliva, L. Forni, A.M. Ezerets, I.E. Mukovozov, A.V. Vishniakov, EPR characterization of $(CeO_2)_{1-y}(La_2CuO_4)_y$ oxide mixtures and their catalytic activity for NO reduction by CO. J. Chem. Soc. Faraday Trans. **94**, 587–592 (1998)

98. S.D. Peter, E. Garbowski, N. Guilhaume, V. Perrichon, M. Primet, Catalytic properties of La_2CuO_4 in the CO + NO reaction. Catal. Lett. **54**, 79–84 (1998)

99. A.A. Galkin, B.G. Kostyuk, V.V. Lunin, M. Poliakoff, Continuous reactions in supercritical water: A new route to La_2CuO_4 with a high surface area and enhanced oxygen mobility. Angew. Chem. Int. Ed. **39**, 2738–2740 (2000)

100. S.N. Torbin, M.N. Danchevskaya, L.F. Martynova, G.P. Muravieva, Role of intermediate solid phase in the process of magnesium and lanthanum aluminates formation in sub- and supercritical water. High Pressure Res. **20**(1–6), 109–119 (2001)

101. M. Emirdag-Eanes, M. Krawiec, J.W. Kolis, Hydrothermal synthesis and structural characterization of $NaLnGeO_4$ (Ln = Ho, Er, Tb, Tm, Yb, Lu) family of lanthanide germinates. J. Chem. Crystallogr. **31**(5), 281–285 (2001)

102. J.B. Gadhe, R.B. Gupta, Hydrogen production by methanol reforming in supercritical water: Catalysis by in-situ-generated copper nanoparticles. Int. J. Hydrogen Energy **32**(13), 2374–2381 (2007)

103. K.J. Ziegler, R.C. Doty, K.P. Johnston, B.A. Korgel, Synthesis of organic monolayer-stabilized copper nanocrystals in supercritical water. J. Am. Chem. Soc. **123**, 7797–7803 (2001)

104. T. Mousavand, S. Takami, M. Umetsu, S. Ohara, T. Adschiri, Supercritical hydrothermal synthesis of organic-inorganic hybrid nanoparticles. J. Mater. Sci. **41**(5), 1445–1448 (2006)

105. T. Adschiri, Supercritical hydrothermal synthesis of organic-inorganic hybrid nanoparticles. Chem. Lett. **36**(10), 1188–1193 (2007)

106. D. Rangappa, S. Ohara, M. Umetsu, T. Naka, T. Adschiri, Synthesis, characterization and organic modification of copper manganese oxide nanocrystals under supercritical water. J. Supercrit. Fluids **44**(3), 441–445 (2008)

107. C.B. Xu, A.S. Teja, Continuous hydrothermal synthesis of iron oxide and PVA-protected iron oxide nanoparticles. J. Supercrit. Fluids **44**(1), 85–91 (2008)

108. X. Wei, G. Xu, Z.H. Ren, C.X. Xu, G. Shen, G.R. Han, PVA-assisted hydrothermal synthesis of $SrTiO_3$ nanoparticles with enhanced photocatalytic activity for degradation of RhB. J. Am. Ceram. Soc. **91**(11), 3795–3799 (2008)

109. J. Yamanaka, S. Mori, Y. Kaneko, Hydrothermal synthesis of vanadium-based layered compound with 1 nm basal spacing. Mater. Trans. **42**(9), 1854–1857 (2001)

110. P.M. Almond, T.E. Albrecht-Schmitt, Hydrothermal synthesis and crystal chemistry of the new strontium uranyl selenites, $Sr[(UO_2)_3(SeO_3)_2O_2]\bullet4H_2O$ and $Sr[UO_2(SeO_3)_2]$. Am. Mineral. **89**(7), 976–980 (2004)

111. J. Otsu, Y. Oshima, New approaches to the preparation of metal or metal oxide particles on the surface of porous materials using supercritical water: Development of supercritical water impregnation method. J. Supercrit. Fluids **33**(1), 61–67 (2005)

112. O. Sawai, Y. Oshima, Mechanism of silver nano-particles formation on alpha-alumina using supercritical water. J. Mater. Sci. **43**(7), 2293–2299 (2008)

113. H. Reveron, C. Elissalde, C. Aymonier, O. Bidault, M. Maglione, F. Cansell, Supercritical fluid route for synthesizing crystalline barium strontium titanate nanoparticles. J. Nanosci. Nanotechnol. **5**(10), 1741–1744 (2005)

114. J. Lu, Y. Hakuta, H. Hayashi, T. Ohashi, Y. Hoshi, K. Sato, M. Nishioka, T. Inoue, S. Hamakawa, Continuous hydrothermal preparation of partially substituted perovskite oxide nanoparticles. Chem. Lett. **36**(10), 1262–1263 (2007)

115. J.F. Lu et al., Preparation of $Ca_{0.8}Sr_{0.2}Ti_{1-x}Fe_xO_{3-\delta}$ (x = 0.1–0.3) nanoparticles using a flow supercritical reaction system. J. Supercrit. Fluids **46**(1), 77–82 (2008)

116. M. Takesue, A. Suino, Y. Hakuta, H. Hayashi, R.L. Smith, Formation mechanism and luminescence appearance of Mn-doped zinc silicate particles synthesized in supercritical water. J. Solid State Chem. **181**(6), 1307–1313 (2008)

117. M. Takesue, A. Suino, K. Shimoyama, Y. Hakuta, H. Hayashi, R.L. Smith, Formation of α- and β-phase Mn-doped zinc silicate in supercritical water and its luminescence properties at Si/(Zn+Mn) ratios from 0.25 to 1.25. J. Cryst. Growth **310**(18), 4185–4189 (2008)

118. M. Takesue, K. Shimoyama, S. Murakami, Y. Hakuta, H. Hayashi, R.L. Smith, Phase formation of Mn-doped zinc silicate in water at high-temperatures and high-pressures. J. Supercrit. Fluids **43**(2), 214–221 (2007)

119. J.Y. Chang, J.J. Chang, B. Lo, S.H. Tzing, Y.C. Ling, Silver nanoparticles spontaneous organize into nanowires and nanobanners in supercritical water. Chem. Phys. Lett. **379**(3–4), 261–267 (2003)

120. Z.Y. Sun, Z.M. Liu, B.X. Han, Y. Wang, J. Du, Z. Xie, G. Han, Fabrication of ruthenium-carbon nanotube nanocomposites in supercritical water. Adv. Mater. **17**(7), 928–932 (2005)

121. O. Sawai, Y. Oshima, Deposition of silver nano-particles on activated carbon using supercritical water. J. Supercrit. Fluids **47**(2), 240–246 (2008)

122. Z. Fang, H. Assaaoudi, H.B. Lin, X.M. Wang, I.S. Butler, J.A. Kozinski, Synthesis of nanocrystalline SnO_2 in supercritical water. J. Nanopart. Res. **9**(4), 683–687 (2007)

123. Z. Fang, H. Assaaoudi, R.I.L. Guthrie, J.A. Kozinski, I.S. Butler, Continuous synthesis of tin and indium oxide nanoparticles in sub- and supercritical water. J. Am. Ceram. Soc. **90**(8), 2367–2371 (2007)

Chapter 4
Nano-Structured Coatings

Abstract Carbide ceramics were coated with a nano-structured carbon film (10 nm~1 μm) that could reduce the friction coefficient by SCW treatment. SCW is very corrosive to a stainless steel made reactor particularly when acids and oxygen are added or formed. At such conditions, the reactor was corroded to form water soluble metal cations that were further hydrolyzed and subsequently dehydrated to fine metal oxide particles. In order to avoid corrosion, the reactor could be pretreated in SCW by adding metal salts [e.g., $Ce(NO_3)_2$] to form a metal oxide (e.g., CeO_2) nano-structured layer on the wall surface. We can also protect the reactor by adding ionic light metal salts (e.g., $NaCl$, Na_2CO_3) in water that will precipitate to fine particles to form a layer on the wall at supercritical conditions. The fine salt particles can be obtained by removing them from supercritical solution after they are precipitated in SCW.

4.1 Carbon Film

Carbide ceramics such as SiC, B_4C, WC and TiC have good wear resistance and find a number of applications because of their high hardness. Better tribological performance can be expected if the ceramic is coated with a carbon film to reduce the friction coefficient. Nano-structured carbon coatings on the surface of SiC and other metal carbides in SCW were achieved [1]. In these experiments, SiC samples were placed in sub- or super-critical water at 573~1123 K and 10~500 MPa (batch reactors). The following reactions were predicted by thermodynamic calculations for the Si–C–H–O system to lead to solid carbon formation [2]:

$$SiC + 2\,H_2O \rightarrow SiO_2 + C + 2\,H_2 \tag{4.1}$$

The critical factor was that as soon as silica was formed, it dissolved in the fluid. Thus reaction (Eq. 4.1) tended to move further to the right, leading to the growth of the carbon coating. Very smooth and uniform carbon films on SiC fibers with thickness of 10 nm~1 μm were obtained and used for interfacial engineering in

Z. Fang, *Rapid Production of Micro- and Nano-particles Using Supercritical Water,* 57
Engineering Materials 30, DOI 10.1007/978-3-642-12987-2_4,
© Springer-Verlag Berlin Heidelberg 2010

ceramic matrix composites, to ensure fiber pull-out and gracious fracture of composites [3–4]. Various carbon allotropes, including diamond, were formed during SCW treatment of SIC. Their structure varied depending on the experimental conditions and SiC precursor [5].

4.2 Oxides Layer

SS reactors are easily corroded in SCW due to high concentration of the ion product (K_w, Fig. 1.9), particularly during SCWO of halogen-containing wastes (e.g., decachlorobiphenyl; $C_{12}Cl_{10}$), where acids are formed [6], e.g., HCl:

$$C_{12}Cl_{10} + 19\ H_2O_2(H_2O + 1/2\ O_2) \rightarrow 12\ CO_2 + 14\ H_2O + 10\ HCl \qquad (4.2)$$

In our previous work during SCWO of decachlorobiphenyl in 316-SS tubular reactors [6], the reactors became severely corroded, resulting in the formation of a large amount of solid material containing hexagonal-, needle-, and rhombic-shaped crystals (Fig. 4.1). The solid phase was definitely from the reactor wall corrosion since analysis by EDX revealed that the majority of elements were O and Fe, and minor elements of Si (traces of Cr) as shown in Fig. 4.2. EDX mapping (Fig. 4.3, parts a and b) showed the presence of O and Fe distributed over all areas of the sample, which is the evidence that the main component in the solid phase were FeO_x. Si and Cr were concentrated in some parts of the sample area (Fig. 4.3, parts c and d), presumably in the form of oxides.

The reactor material, 316-SS, is an Fe-based alloy with an approximate chemical composition of 65% Fe, 17% Cr, 12% Ni, 2.5% Mo, 2.0% Mn, 1.0% Si, and other trace elements. Metal (Me) was corroded to $MeCl_x$ according to Eq. 4.3, and crystal metal oxides ($MeO_{x/2}$) were formed according to Eqs. 4.4 and 2.2:

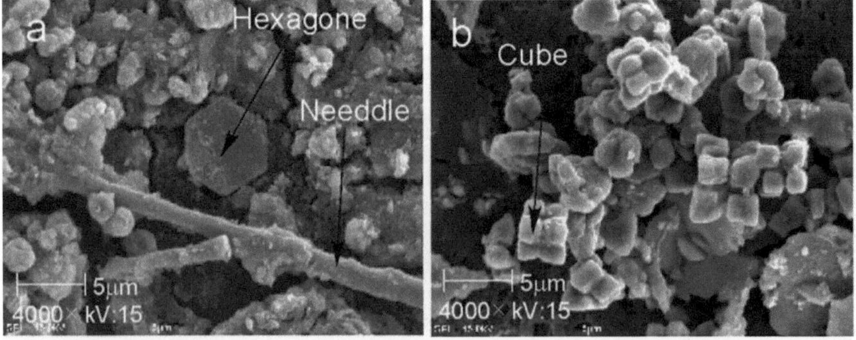

Fig. 4.1 SEM micrographs of the solid residues after SCWO of decachlorobiphenyl in 316-SS batch reactors at 723 K and 30 MPa for 600 s with 93% excessive O_2 according to Eq. 4.2. Reprinted with permission from [6]. Copyright © 2004, American Chemical Society

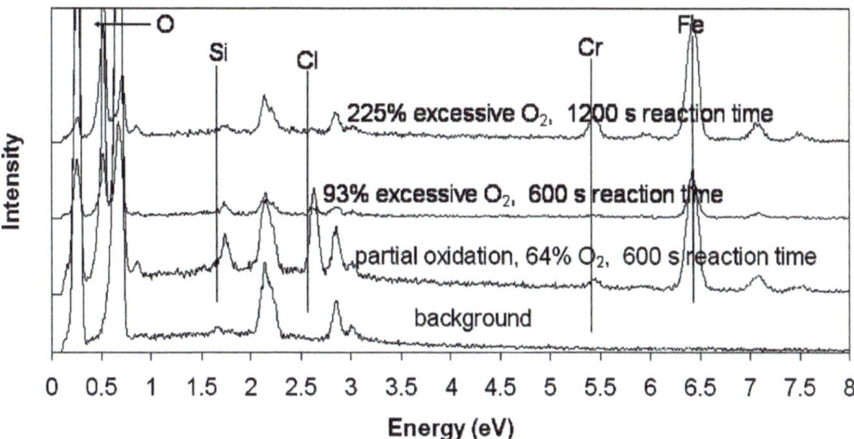

Fig. 4.2 EDX spectra of the solid residues after SCWO of decachlorobiphenyl in 316-SS batch reactors at 723 K and 30 MPa. Reprinted with permission from [6]. Copyright © 2004, American Chemical Society

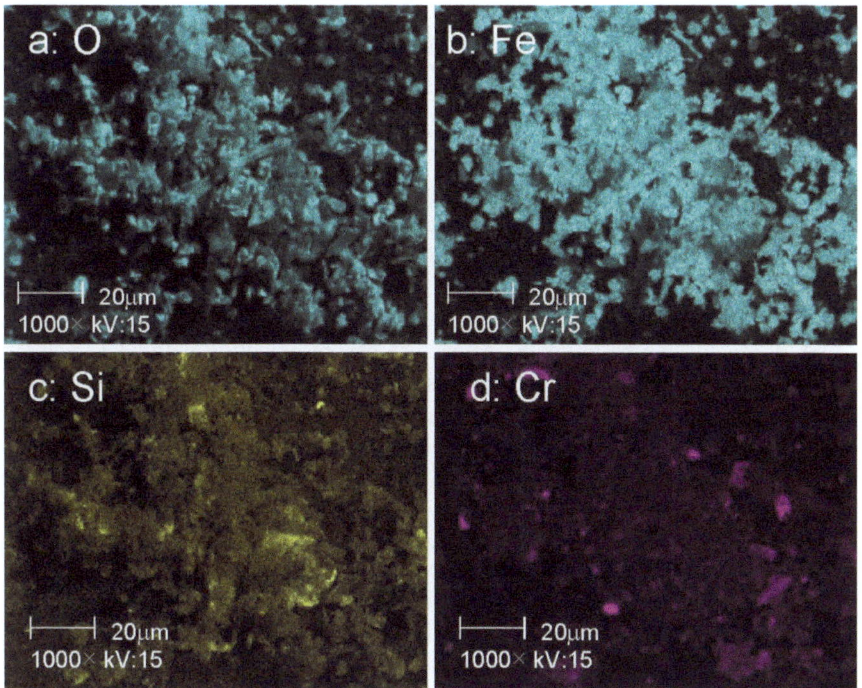

Fig. 4.3 EDX mapping of the solid residues after SCWO of decachlorobiphenyl in 316-SS batch reactors at 723 K and 30 MPa for 600 s with 93% excessive O_2 according to Eq. 4.2. Reprinted with permission from [6]. Copyright © 2004, American Chemical Society

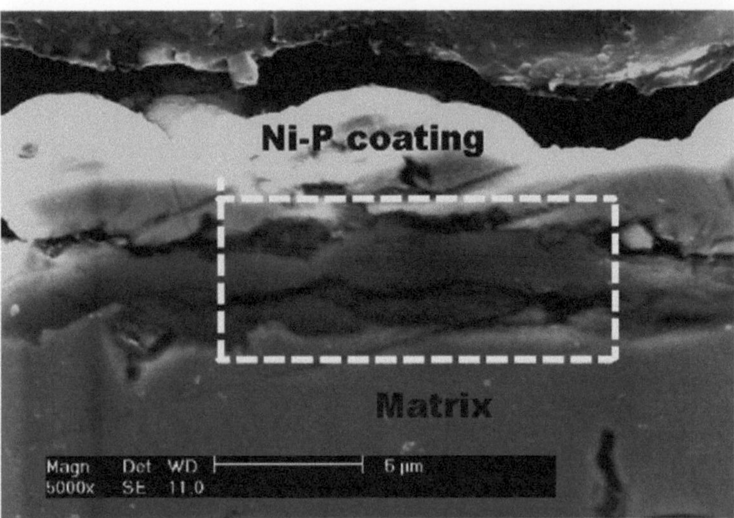

Fig. 4.4 Cross section of a continuous oxide film formed at 773 K and 24 MPa for 100 h in SCW with 2% H_2O_2 (Ni–P coating is for protecting the film for subsequent analysis). Reprinted with permission from [7]. Copyright © 2007, Elsevier

$$Me + x\ HCl \rightarrow MCl_x + x/2\ H_2 \qquad (4.3)$$

$$MeCl_x + x\ H_2O \rightarrow Me(OH)_x + x\ HCl \qquad (4.4)$$

Gao et al. [7] studied oxide films grown on 316 SS sheet exposed to H_2O_2-containing SCW and a duplex-layer structure was identified in the oxide films. The loose outer layer was rich in Fe, while the compact inner layer was rich in Cr and Fe. Ni enrichment was observed at the interface between the metal matrix and oxides (Fig. 4.4).

SCW reactor can be protected before use by pre-oxidation with heat treatment in air at high temperatures or by exposure to SCW. However, preoxidation can also result in degraded performance if not performed carefully or under the right conditions [8]. An investigation was conducted for the corrosion behavior of high-temperature alloys such as alloy 602CA (UNS N06602) and alloy G-30 (UNS N06030), that were exposed to SCW with oxygen and Cl-contained compounds at 40 MPa and 693 K for 24 h. The corrosion rate of alloy 602CA was reduced by a factor of three if preoxidated at 1,173 K prior to SCWO exposure. Alloy G-30 showed good corrosion behavior due to the formation of a protective oxide layer ($\sim$12 µm) during the exposure [9]. Corrosion could be well protected under SCW oxidation conditions by the formation of a Ce oxide nano-structured layer on the surface of the tube [10] or by formation of α-Fe_2O_3 and Fe_3O_4 nanoparticle layer. In other work, potassium titanate (K_2TiO_3) fibers were grown on the surface of a tubular reactor with a titanium liner that was formed by reaction of Ti from the liner with $KHCO_3$ in SCW [11]. The fiber layer may protect the reactor from corrosion.

4.3 Ligh Metal Salt Layer

At supercritical conditions, water becomes weakly-polar solvent, solubility of ionic salts decreases significantly. Many light metal salts that have high solubility in liquid water have extremely low solubility in SCW (Fig. 4.5) [12]. For example, NaCl solubility is about 37 wt% at 573 K but only about 120 ppm at 823 K and 25 MPa, dropped by 99.97% [13]; $CaCl_2$ has a maximum solubility of 70 wt% at subcritical temperatures, dropped to only 3 ppm at 773 K and 25 MPa [14]. The solubility of other salts and oxides in SCW, such as nitrates ($LiNO_3$, $NaNO_3$, KNO_3) [15], lead (II) and copper (II) oxides [16] has also been studied. Therefore, light metal salts will precipitate as fine particles on the wall of a reactor that can protect the reactor from corrosion. Muthukumaran and Gupta [17] found that a 0.25 wt% loading of sodium carbonate can give a particle surface area that is about 133 times the surface area of the reactor wall assuming average particle size to be 1 μm. This represents an extremely large corrosion protection in view of adsorption sites for the corrosive species.

In SCWO of decachlorobiphenyl, we found that adding Na_2CO_3 greatly reduced the corrosion and promoted the destruction rate [6]. Because Na_2CO_3 was used as a neutralizing agent to remove HCl formed in Eq. (4.2):

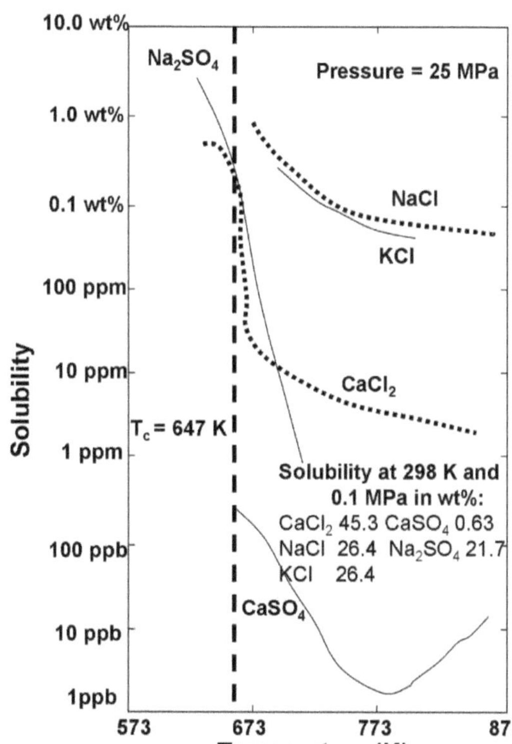

Fig. 4.5 Salt solubility in supercritical water at 25 MPa. Reprinted with permission from [12]. Copyright © 1993, American Chemical Society

$$10\,HCl + 5\,Na_2CO_3 \rightarrow 10\,NaCl\downarrow + 5\,CO_2 + 5\,H_2O \qquad (4.5)$$

Adding Na_2CO_3 also resulted in formation of a salt $\{NaCl + Na_2CO_3\}$ particle layer that can protect the reactor. We can also obtain fine salt particles by removing them from supercritical solution after they are precipitated upon heating to supercritical region.

References

1. Y. Gogotsi, Nanostructured carbon coatings. NATO Sci. Ser. 3 High Technol. **78**, 25–40 (2000)
2. N.S. Jacobson, Y.G. Gogotsi, M. Yoshimura, Thermodynamic and experimental study of carbon formation on carbides under hydrothermal conditions. J. Mater. Chem. **5**, 595–601 (1995)
3. Y.G. Gogotsi, M. Yoshimura, Formation of carbon films on carbides under hydrothermal conditions. Nature **367**, 628–630 (1994)
4. Y.G. Gogotsi, S. Welz, J. Daghfal, M.J. McNallan, I.D. Jeon, K.G. Nickel, T. Kraft, Formation of carbon coatings on SiC fibers by selective etching in halogens and supercritical water. Ceram. Eng. Sci. Proc. **19**, 87–94 (1998)
5. Y.G. Gogotsi, P. Kofstad, M. Yoshimura, K.G. Nickel, Formation of sp^3-bonded carbon upon hydrothermal treatment of SiC. Diamond Relat. Mater. **5**, 151–162 (1996)
6. Z. Fang, S. Xu, I.S. Butler, R.L. Smith Jr., J.A. Kozinski, Destruction of decachlorobiphenyl using supercritical water oxidation. Energy & Fuels **18**(5), 1257–1265 (2004)
7. X. Gao, X. Wu, Z. Zhang, H. Guan, E. Han, Characterization of oxide films grown on 316L stainless steel exposed to H_2O_2-containing supercritical water. J. Supercrit. Fluids **42**(1), 157–163 (2007)
8. P.A. Marrone, G.T. Hong, Corrosion control methods in supercritical water oxidation and gasification processes. J. Supercrit. Fluids **51**(2), 157–163 (2009)
9. J. Konys, S. Fodi, J. Hausselt, H. Schmidt, V. Casal, Corrosion of high-temperature alloys in chloride-containing supercritical water oxidation systems. Corrosion **55**(1), 45–51 (1999)
10. V.S. Rao, H.S. Kwon, Reactor corrosion in ceria production by hydrothermal synthesis under supercritical conditions. Corrosion **63**(4), 359–365 (2007)
11. W. Habicht, N. Boukis, G. Franz, O. Walter, E. Dinjus, Exploring hydrothermally grown potassium titanate fibers by STEM-in-SEM/EDX and XRD. Microsc. Microanal. **12**(4), 322–326 (2006)
12. J.W. Tester, H.R. Holgate, F.J. Armellini, P.A. Webley, W.R. Killilea, G.T. Hong, H.E. Barner, Supercritical water oxidation technology: Process development and fundamental research. ACS Symp. Ser. **518**, 35–76 (1993)
13. K.S. Pitzer, R.T. Pabalan, Thermodynamics of sodium chloride in steam. Geochim. Cosmochim. Acta **50**, 1445–1454 (1986)
14. I. Martynova, Solubility of inorganic compounds in subcritical and supercritical water. Int. Corros. Conf. Ser. **NACE-4** (High Temp. High Pressure Electrochem. Aqueous Solutions, Conf.) 131–138 (1976)
15. P. Dell'Orco, H. Eaton, T. Reynolds, S. Buelow, The solubility of 1:1 nitrate electrolytes in supercritical water. J. Supercrit. Fluids **8**, 217–227 (1995)
16. K. Sue, Y. Hakuta, R.L. Smith Jr., T. Adschiri, K. Arai, Solubility of lead(II) oxide and copper(II) oxide in subcritical and supercritical water. J. Chem. Eng. Data **44**, 1422–1426 (1999)
17. P. Muthukumaran, R.B. Gupta, Sodium-carbonate-assisted supercritical water oxidation of chlorinated waste. Ind. Eng. Chem. Res. **39**, 4555–4563 (2000)

Chapter 5
Other Materials Synthesis

Abstract Diamond was synthesized from graphite in SCW at high temperatures and ultra-high pressures ($\sim$3,273 K, 10 GPa) for long time (e.g., 24 h). Many inorganic phosphates were produced that have important industrial uses, e.g., as catalysts, ion-exchange materials, solid electrolytes for batteries and synthetic replacements for bones and teeth. Submicron particles of $LiFePO_4$ were produced in SCW from $\{FeSO_4 + o - H_3PO_4 + LiOH\}$ solution in batch and flow reactors. Much smaller and more uniform particles were found in rapid continuous synthesis than in batch hydrothermal synthesis. Micron-sized $KCo_3Fe(PO_4)_3$ and $Mg_{3.5}H_2(PO_4)_3$ crystals were also produced in batch reactors. At the same time, amorphous nanoparticles of $K_xCo_yFe_zPO_4$ and $Fe_{1-y}K_yPO_4$ were obtained.

5.1 Diamond

Graphite can be converted to diamond in SCW at very high temperatures and ultra-high pressures ($\sim$3,273 K, 10 GPa) [1]. Kumar et al. and Yamaoka et al. [2–3] synthesized small octahedral diamond crystals ($< 10 \, \mu m$) from graphite and oxalic acid $(COOH)_2$ in a platinum-sealed capsule at 7.7 GPa and 1,573$\sim$1,773 K for 24 h (Fig. 5.1). Diamond synthesis from graphite in the presence of water and SiO_2 was also studied at 1,723$\sim$1,773 K and 8$\sim$8.5 GPa [4]. Diamond crystallization in the Si–C–O–H system required more reduced media than its counterpart synthesized in Mg–C–O–H and Ca–Mg–C–O–H systems at similar pressures, temperatures and reaction times. Formation of diamond under less extreme SCW conditions (e.g., 140 MPa and 1,073 K) has been reported [5, 6]. Szymanski et al. [6] proposed a metastable diamond nucleation and growth region at low pressure SCW [6].

Gogotsi et al. [7] studied diamond powders and crystals treatment in SCW at 923$\sim$1,123 K and pressures up to 500 MPa. It is found that a noticeable interaction

Z. Fang, *Rapid Production of Micro- and Nano-particles Using Supercritical Water*,
Engineering Materials 30, DOI 10.1007/978-3-642-12987-2_5,
© Springer-Verlag Berlin Heidelberg 2010

Fig. 5.1 SEM micrograph of diamond crystals spontaneously nucleated from high pressure and high temperature water conditions at 7.7 GPa and 1,773 K for 24 h. Reprinted with permission from [3]. Copyright © 2000, Elsevier

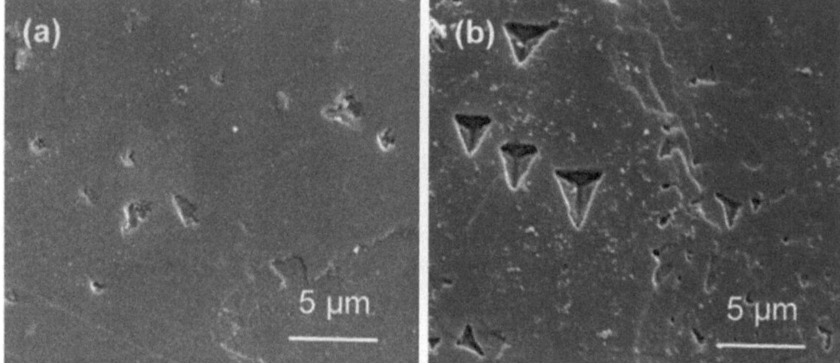

Fig. 5.2 SEM micrographs of the surface of diamond single crystals (**a**) as received and (**b**) after SCW treatment for 15 h at 1,073 K and 100 MPa. Reprinted with permission from [7]. Copyright © 1998, Elsevier

of fine grain diamond powders with H_2O started at 973 K, etching pits appeared on the single crystal diamond surface at 1,073~1,123 K (Fig. 5.2).

5.2 Inorganic Phosphates

Inorganic phosphates have numerous important industrial uses, e.g., as catalysts, ion-exchange materials, solid electrolytes for batteries, materials in linear and non-linear optical components, as chelating agents, synthetic replacements for bones and teeth, phosphors, detergents and fertilizers. They were also synthesized in SCW [8–11]. Submicron particles of $LiFePO_4$ were obtained in SCW from $\{FeSO_4 + o - -H_3PO_4 + LiOH\}$ solution in batch and flow reactors according to the reaction [11]:

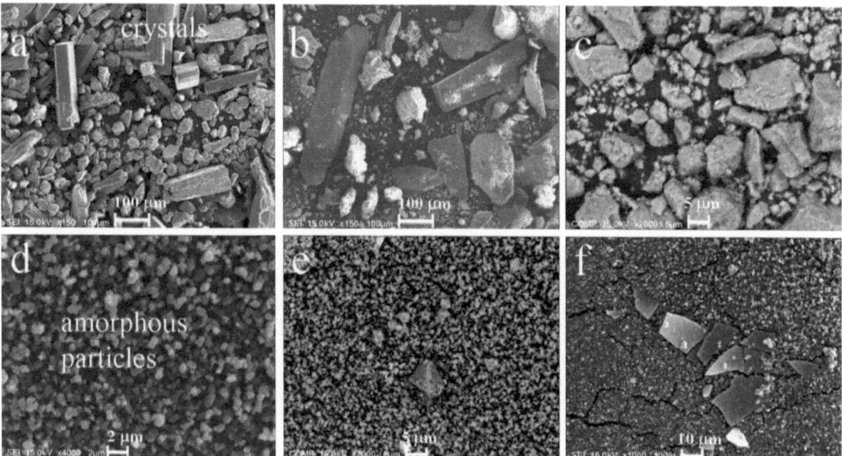

Fig. 5.3 SEM micrographs of micro-sized crystals of $KCo_3Fe(PO_4)_3$ and amorphous particles synthesized in SCW at 723 K, 32 MPa and 2 h in 316-SS batch reactors

$$FeSO_4 + H_3PO_4 + 3\,LiOH \rightarrow LiFePO_4\downarrow + Li_2SO_4 + 3\,H_2O \qquad (5.1)$$

Much smaller and more uniform particles were found in continuous hydrothermal synthesis than in batch hydrothermal synthesis. Nanosized (28-nm sphere) magnesium substituted calcium phosphate bioceramics was synthesized in subcritical water at 548 K continuously [12].

We synthesized crystals of a new potassium cobalt(II) iron(II) triorthophosphate salt, $KCo_3Fe(PO_4)_3$ as well as amorphous nanoparticles from $\{CoCl_2 + K_4P_2O_7 + HCl\}$ (Fig. 5.3) in a 316-SS batch reactor at 673~723 K and 25~32 MPa [9].

The crystals and amorphous nanoparticles had the same elemental compositions due to similar EDX spectra (Fig. 5.4a, b). The Fe(II) in the new orthophosphate salt should be derived from leaching of iron from the 316-SS reactor according to reaction in Eq. (4.3). The salt crystallized in the orthorhombic space group Pnnm, $Z = 4$, with unit cell parameters as follows: $a = 9.672(1)$ Å, $b = 16.457(2)$ Å, $c = 6.2004(8)$ Å, $V = 986.97$ Å^3 [9]. Figure 5.5 gives the crystal morphology changed with reaction temperature and time.

At similar conditions, using the same type of batch reactors, we synthesized crystals of a new magnesium hydrogen orthophosphate salt, $Mg_{3.5}H_2(PO_4)_3$, as well as amorphous nanoparticles, non-Mg containing phosphate material, $Fe_{1-y}K_yPO_4(0 < y < 1)$ from $\{MgCl_2 + K_4P_2O_7 + HCl\}$ solution (Figs. 5.6 and 5.7) [10]. The salt crystallized in the triclinic space group Pl, $Z = 2$ with the following unit-cell parameters: $a = 6.438(1)$, $b = 7.856(1)$, $c = 9.438(1)$ Å; $\alpha = 104.57(1)$, $\beta = 108.61(1)$, $\gamma = 101.28(1)°$, $V = 739.99$ Å^3. EDX spectra and optical microscopic images of the products are given in Figs. 5.7, 5.8, and 5.9.

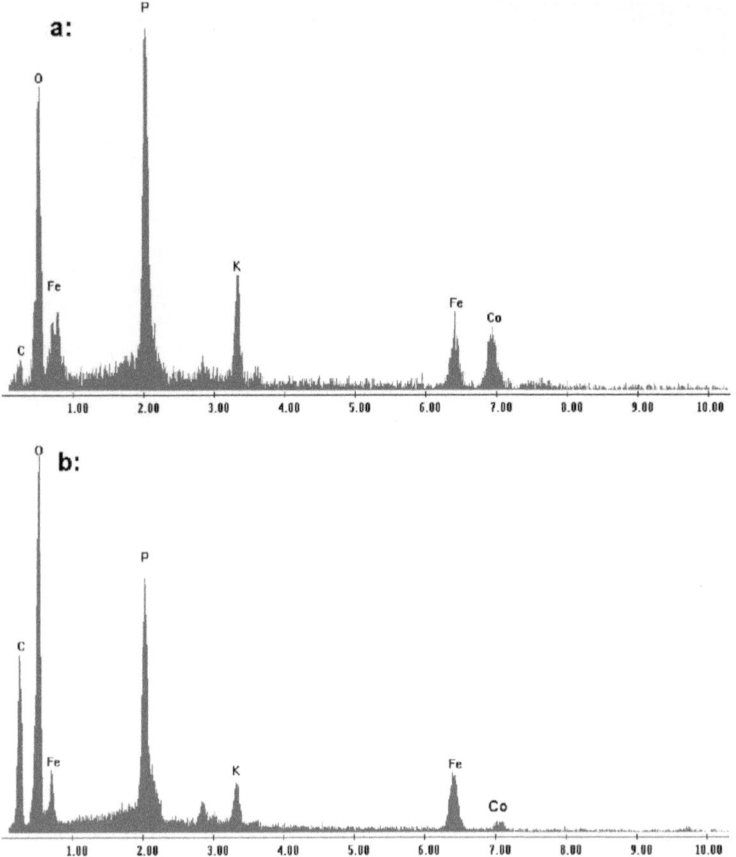

Fig. 5.4 EDX spectra for (**a**) a selected crystal and (**b**) amorphous particles in Fig. 5.3

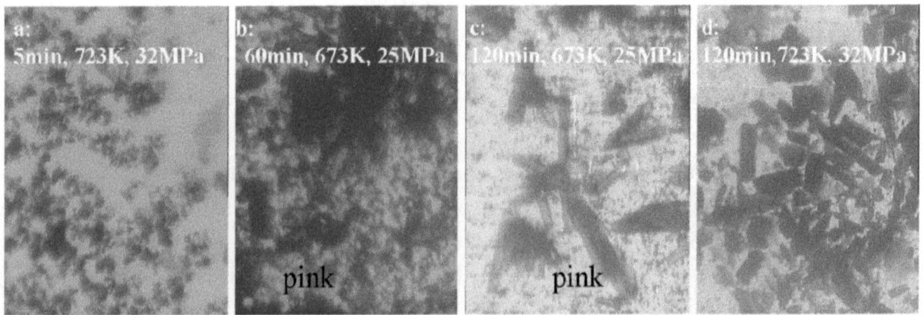

Fig. 5.5 Optical microscopic images ($\times 100$) of $KCo_3Fe(PO_4)_3$ crystals obtained in SCW at 673~723 K, 25~32 MPa for 30~120 min reaction time

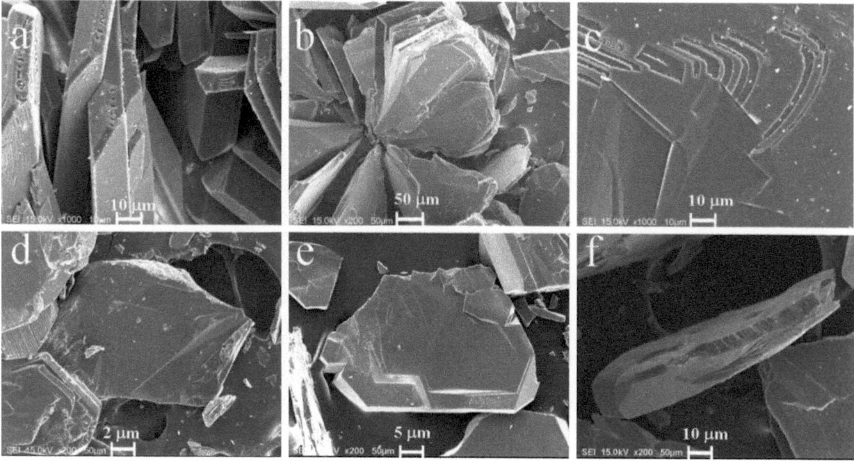

Fig. 5.6   SEM micrographs of micro-sized crystals of $Mg_{3.5}H_2(PO_4)_3$ synthesized in SCW

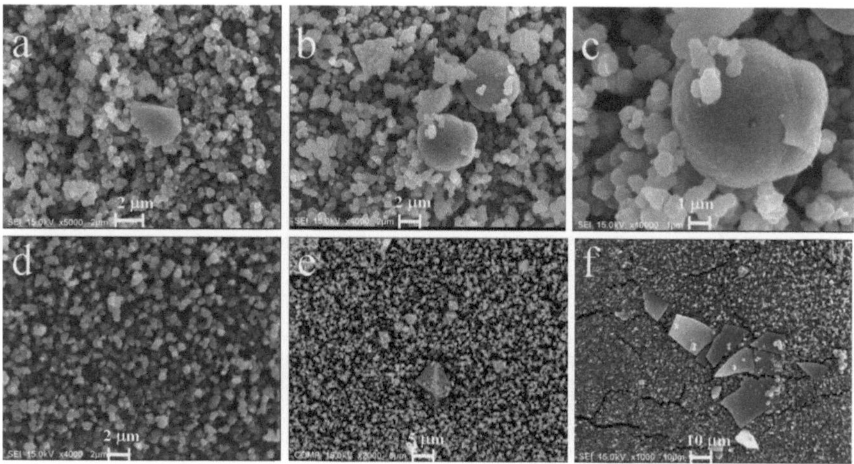

Fig. 5.7   SEM micrographs of amorphous particles of $Fe_{1-y}K_yPO_4$ ($0 < y < 1$) synthesized in SCW

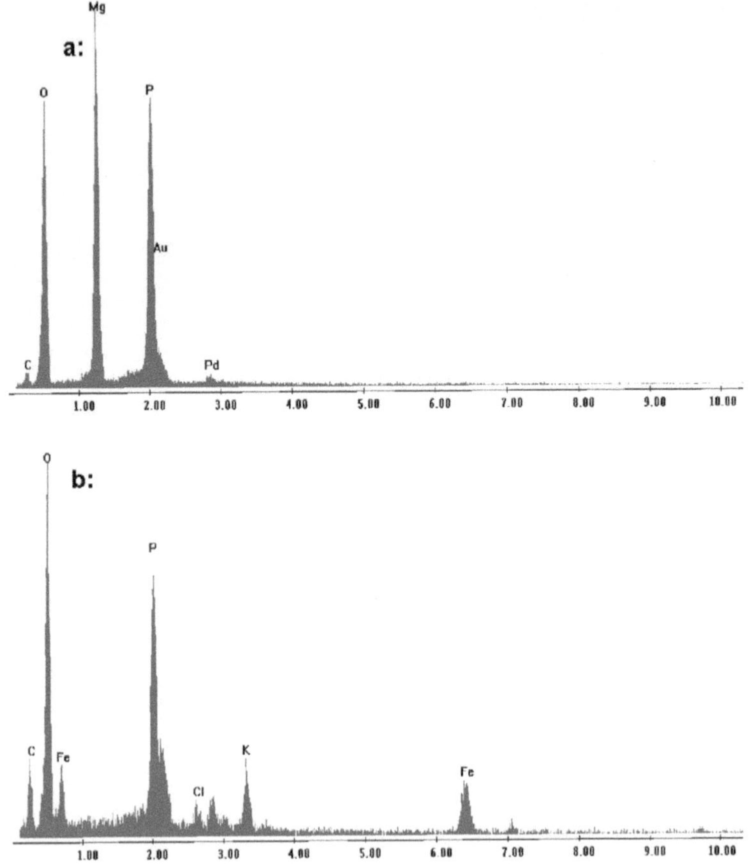

Fig. 5.8 EDX spectra for **(a)** a selected crystal in Fig. 5.6 and **(b)** amorphous particles in Fig. 5.7

Fig. 5.9 Optical microscopic images ($\times$100) of $Mg_{3.5}H_2(PO_4)_3$ crystals obtained in SCW at 673$\sim$723 K, 25$\sim$32 MPa for 30$\sim$120 min reaction time

References

1. T.L. Brown, H.E. LeMay, B.E. Bursten, *Chemistry: The Central Science*, 8th edn. (Prentice-Hall, Upper Saddle River, NJ, 1999)
2. M.D. S. Kumar, M. Akaishi, S. Yamaoka, Formation of diamond from supercritical H_2O–CO_2 fluid at high pressure and high temperature. J. Cryst. Growth **213**, 203–206 (2000)
3. S. Yamaoka, M.D.S. Kumar, M. Akaishi, H. Kanda, Reaction between carbon and water under diamond-stabl high pressure and high temperature conditions. Diamond Relat. Mater. **9**(8), 1480–1486 (2000)
4. L.F. Dobrzhinetskaya, H.W. Green, Diamond synthesis from graphite in the presence of water and SiO_2: Implications for diamond formation in quartzites from Kazakhstan. Int. Geol. Rev. **49**(5), 389–400 (2007)
5. X.Z. Zhao, R. Roy, K.A. Cherian, A. Badzian, Hydrothermal growth of diamond in metal-C–H_2O systems. Nature **385**, 513–515 (1997)
6. A. Szymanski, E. Abgarowicz, A. Bakon, A. Niedbalska, R. Salacinski, J. Sentek, Diamond formed at low pressures and temperatures through liquid-phase hydrothermal synthesis. Diamond Relat. Mater. **4**, 234–235 (1995)
7. Y. Gogotsi, T. Kraft, K.G. Nickel, M.E. Zvanut, Hydrothermal behavior of diamond. Diamond Relat. Mater. **7**, 1459–1465 (1998)
8. J. Lee, A.S. Teja, Characteristics of lithium iron phosphate ($LiFePO_4$) particles synthesized in subcritical and supercritical water. J. Supercrit. Fluids **35**(1), 83–90 (2005)
9. H. Assaaoudi, Z. Fang, D.H. Ryan, I.S. Butler, J.A. Kozinski, Hydrothermal synthesis, crystal structure, and vibrational and Mössbauer spectra of a new tricationic orthophosphate – $KCo_3Fe(PO_4)_3$. Can. J. Chem. **84**(2), 124–133 (2006)
10. H. Assaaoudi, Z. Fang, I.S. Butler, D.H. Ryan, J.A. Kozinski, Characterization of a new magnesium hydrogen orthophosphate salt, $Mg_{3.5}H_2(PO_4)_3$, synthesized in supercritical water. Solid State Sci. **9**(5), 385–393 (2007)
11. C.B. Xu, J. Lee, A.S. Teja, Continuous hydrothermal synthesis of lithium iron phosphate particles in subcritical and supercritical water. J. Supercrit. Fluids **44**(1), 92–97 (2008)
12. A.A. Chaudhry, J. Goodall, M. Vickers, J.K. Cockcroft, I. Rehman, J.C. Knowles, J.A. Darr, Synthesis and characterisation of magnesium substituted calcium phosphate bioceramic nanoparticles made via continuous hydrothermal flow synthesis. J. Mater. Chem. **18**(48), 5900–5908 (2008)

Chapter 6
Fine Organics Particles by Precipitation of Solute

Abstract It was found that non-polar gases (O_2, N_2, CO_2, H_2, He, Ne, Ar, Kr, Xe), aromatics (e.g, benzene, naphthalene) and hydrocarbons (up to C_{36}) could completely dissolve in SCW. Adding salts shifted the homogeneous phase of {water + hydrocarbons} systems to higher temperatures. For example, the addition of only 0.53-mol% NaCl shifted the homogeneous phase of {water + CH_4} to higher temperatures by more than 100 K. Therefore, fine particle can be produced by reducing pressure, temperature and adding salts to precipitate solute from supercritical homogeneous solution. Using DAC techniques, naphthalene, polyethylene terephthalate, nylon, cellulose and wood were completely solubilized in SubCW and SCW, and fine particles were subsequently precipitated by rapid cooling. Nanoparticles were also produced by carbonization of glucose in SubCW and SCW in DAC.

Since SCW has high solubility to dissolve non-polar organics, similar to the RESS process in Chap. 1 (Fig. 1.3), organic particles (e.g., polymers and biomass) can be produced via precipitation of solutes from supercritical solutions after organics are dissolved initially. Particles can also be produced through reaction or carbonization of organics in water phase. In this chapter, DAC technique was used for the study of particles production by reaction and precipitation of solutes from sub- and super-critical water solutions.

6.1 Phase Behavior of Aqueous Systems with Gases and Organics

It is important to know whether and at what conditions organics will dissolve in SCW for the production of fine particles by precipitation of solute. Phase behavior of binary systems of water with gas, alkanes and aromatics, as well as ternary systems of {water + alkanes + salts} up to the supercritical region is introduced and discussed in this section.

Z. Fang, *Rapid Production of Micro- and Nano-particles Using Supercritical Water*,   71
Engineering Materials 30, DOI 10.1007/978-3-642-12987-2_6,

6.1.1 {Non-Polar Gases + Water} Systems

A number of binary aqueous systems with non-polar gases have been studied. These systems have usually interrupted critical curves (CC), the upper branch of which begins at the CP of pure water and then extends into the (T, P, x) space (Fig. 6.1; x is mole fraction of water) [1]. Figure 6.2 shows a typical case for these systems [2]. A three-dimensional (P, T, x) diagram is given in Fig. 6.2a and the CC has a shallow temperature minimum.

6.1.1.1 Binary Aqueous Systems with O_2, N_2 and CO_2

Formation of a homogeneous phase of {O_2 + H_2O} system in SCW is essential to the SCWO process, where O_2 can be pure or from air. A comprehensive knowledge of the water-nitrogen system is needed for an eventual description of the water-air

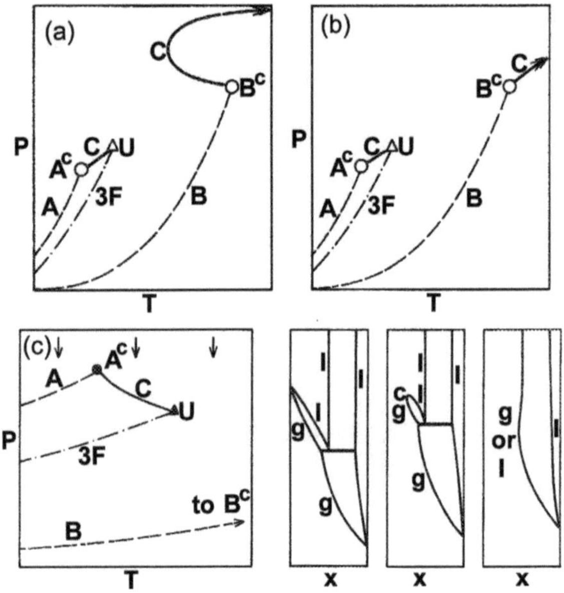

Fig. 6.1 Two-component (*A* and *B*) systems with interrupted critical curves. (**a**) having a temperature minimum, (**b**) proceeding directly to higher temperatures and (**c**) projections for the three-phase (3F) equilibria. g – gas, l – liquid, x – mole fraction; c – critical (point or curve), U (UCEP) – upper critical end point. Reprinted with permission from [1]. Copyright © 1985, Elsevier

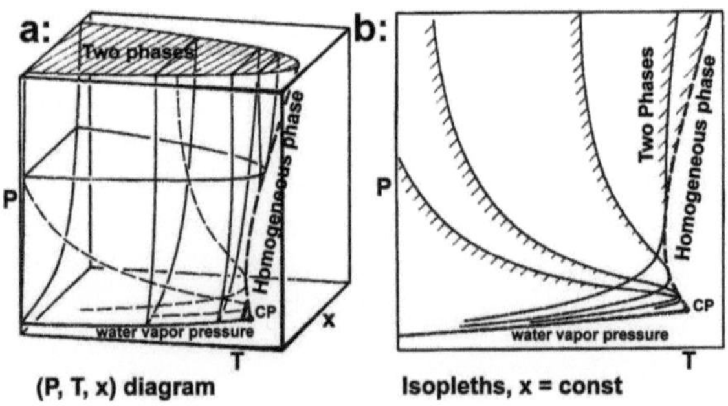

Fig. 6.2 Schematic phase diagram for the system of {xH_2O + (1−x)(non-polar gas)}. x: mole fraction of water; hatched area: two-phase region; CP: critical point of water. Reprinted with permission from [2]. Copyright © 1987, Elsevier

Fig. 6.3 Phase diagram (p, T, x) for the system of $\{xH_2O + (1-x)CO_2\}$. Reprinted with permission from [5]. Copyright © 1963, Oldenbourg

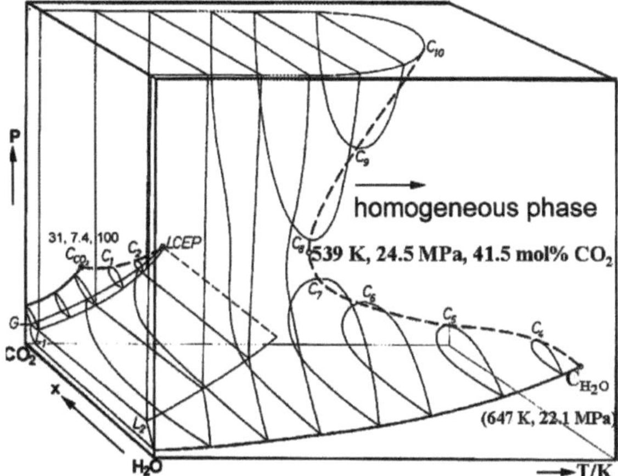

system. Both aqueous systems with N_2 and O_2 have similar behavior. The CC, which begins at the CP of water (647 K), has a slight temperature minimum around 640/639 K at 75 MPa and reaches 250 MPa at 663/659 K for O_2/N_2 [3–4]. An important binary aqueous system $\{H_2O + CO_2\}$ is given in Fig. 6.3 [5]. The temperature minimum of the CC shifts to around 539 K {vs. 640 K for $[O_2 + H_2O]$ system} by lowering 101 K temperature at 24.5 MPa and CO_2 molar fraction of 41.5%.

6.1.1.2 Binary Aqueous Systems with Noble Gases

The CC of {water + xenon} system has been determined experimentally up to 623 K and 300 MPa. It has a temperature minimum around 606 K at 80 MPa [6]. For the system of {water + argon}, the CC has a temperature minimum around 637.5 K at 84 MPa and 73 mol% H_2O [7], which is very close to 639 K for water-nitrogen system. The temperature minimum for $\{xH_2O + (1-x) Kr\}$ system is 631 K and 55 MPa [8]. The rare examples of {gas + gas} immiscibility are found for both $\{H_2O + He\}$ and $\{H_2O + Ne\}$ systems, where the CCs proceed directly form the CP of water directly to higher temperatures and pressures without passing through a temperature minimum [8–9]. The trend for {five noble gases (He, Ne, Ar, Kr, Xe) + water} systems is that the CCs shift to higher temperatures as atomic weight decreases [10].

6.1.2 *{Hydrocarbons + Water} Systems*

6.1.2.1 $\{C_nH_{2n+2} + H_2O\}$ Systems

The $\{H_2 + H_2O\}$ system ($n = 0$), for 0.5 to 90 mol-% H_2 at temperatures up to 713 K and pressures up to 250 MPa, is also found to exhibit so called "gas-gas immiscibility" as the systems of $\{H_2O + He\}$ and $\{H_2O + Ne\}$ [11]. Temperature

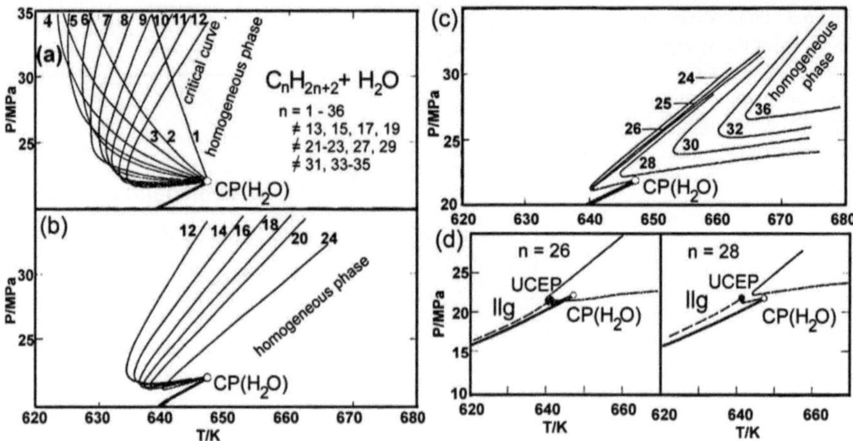

Fig. 6.4 The high-pressure branches of critical curves of {n-alkanes + water} systems (**a–c**) and barotropic effect (**d**). Reprinted with permission from [13]. Copyright © 1990, Elsevier

at this CC changes little from the CP of water as pressure goes up to 250 MPa (e.g., $T_c = 654$ K at $P_c = 252$ MPa for 39 mol-% H_2). The CC of {$CH_4 + H_2O$} system has a temperature minimum and rises steeply to high pressure after passing the minimum point [12, 2].

Brunner [13] summarized phase behavior of 23 n-alkanes (up to $n = 36$) in SCW. All CCs are interrupted and show (gas + gas) equilibria of the second kind (No continuous critical line joining the critical points of the 2 pure components mixtures, which have very different gas-liquid critical temperatures [14]).

The critical temperatures of alkanes are covered from T_c (CH_4) = 190.6 K to T_c ($C_{36}H_{74}$) = 894 K. The investigated alkanes cover volatility regions from higher to lower temperature than that of water (T_c = 647 K). The upper branch of the interrupted CC (Fig. 6.4) with $n = 1\sim26$, which starts from the CP of water, initially passes through temperature minimum, and rises steeply to high pressures as the temperatures rise again. For all mixtures with $n = 7\sim26$, this branch of the CC additionally also passes though a pressure minimum between the CP of water and the temperature minimum. With the carbon number n of the alkanes rising further, the temperature of the UCEP (upper critical end point) rises further. A barotropic effect with reversal of the liquid phases occurs on the llg (liquid-liquid-gas) three-phase line at $n = 28$ (Fig. 6.4d); the consequence is that the water-rich liquid phase becomes identical with the gas phase at UCEP. This means that the low branch of CC which starts from the UCEP, now no longer ends at the CP of the alkane but at the CP of water. The barotropic effect was found with all mixtures with $n \geq 28$, whereas with all mixtures with $n < 26$ no barotropic effect exists on the llg three-phase line.

6.1.2.2 {Aromatics + H_2O} Systems

The phase equilibria of binary aqueous solutions were measured for benzene with or without added salts (KCl, KI), alky-substituted benzenes (methyl-benzene,

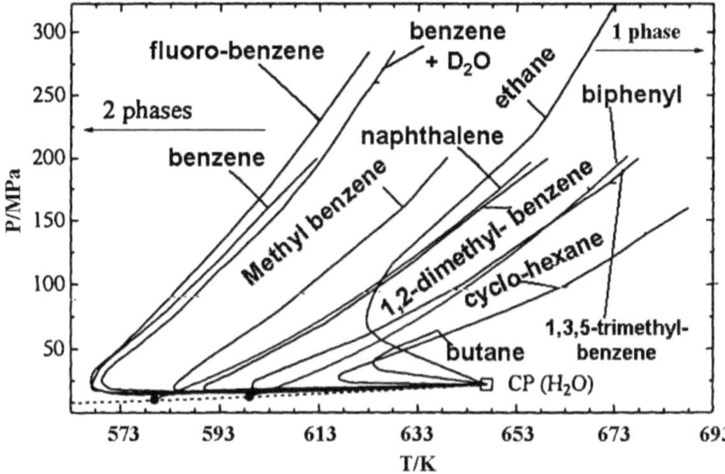

Fig. 6.5 Critical curves for the binary aqueous systems of aromatics. Reprinted with permission from [15]. Copyright © 1969, the Deutsche Bunsen-Gesellschaft

1,2-dimethyl-benzene, 1,3,5-trimethybenzene, fluoro-benzene), biphenyl and naphthalene (Fig. 6.5, other non-aromatics also included) over the temperature range of 533~693 K and up to a maximum pressure of 300 MPa [15–16]. In Fig. 6.5, all curves have a temperature minimum, which means the aromatics can completely dissolve in SubCW, e.g., naphthalene and water are completely miscible in the liquid state above 583 K. Addition of salts shifts the single homogeneous phase of {benzene + water} to higher temperature region [15].

6.1.3 Ternary Systems of {Water + Hydrocarbons + Salts}

Study of the influence of addition of inorganic salts to aqueous binary systems is important to know their phase behavior for the production of fine particles by precipitation.

6.1.3.1 Ternary Systems of {H₂O + CH₄ + Salts}

For the water-methane-salt systems, stability range of different phases is also of importance for exploration and exploitation of natural gas deposits. It is found that in comparison with the binary H₂O–CH₄ system, the addition of only 0.53-mol% NaCl (relative to water) shifts the range of partial immiscibility to higher temperatures by more than 100 K (for example from 610 to 715 K at 100 MPa for 51-mol% CH₄). This shift or "*salting out*" also occurs for the {benzene + water} system mentioned in the above Sect. 6.1.2.2 [15]. Calcium chloride is used to study the influence of a comparable salt with a bivalent cation. Similar "*salting out*" effect is also found by adding CaCl₂ to the system H₂O–CH₄.

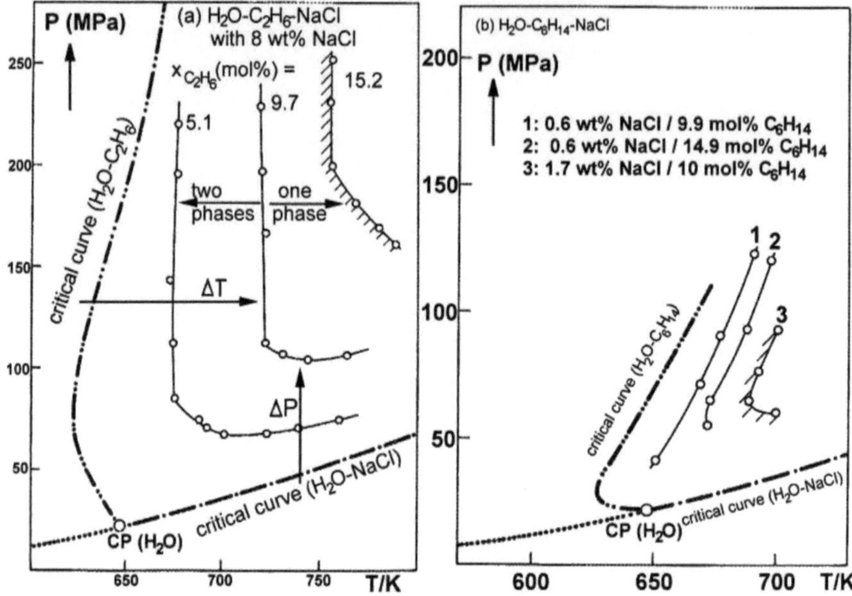

Fig. 6.6 Pressure-temperature diagram of the experimental isopleths for ternary system (**a**) H₂O–C₂H₆–NaCl and (**b**) H₂O–C₆H₁₄–NaCl. Reprinted with permission from [18]. Copyright © 1990, the Deutsche Bunsen-Gesellschaft

6.1.3.2 Ternary System of {H₂O + CO₂ + NaCl}

The system {H₂O + CO₂ + NaCl} has considerable importance for geochemistry (e.g., fluid inclusions in minerals) and also for SCWO process. Gehrig et al. [17] has obtained 20 isopleths with CO_2-concentrations between 0.2 and 85-mol% at 6 wt% NaCl (relative to water). For the isopleth in the water-rich region (4-mol% CO_2), the salt effect is clear but not very pronounced. For a mixture with 48-mol% CO_2, the NaCl addition shifts the two-phase region by about 100 K to higher temperatures.

6.1.3.3 Ternary System of {H₂O + alkane + NaCl}

The phase equilibria of the ternary systems of H₂O–C₂H₆–NaCl and H₂O–C₆H₁₄–NaCl were studied [18]. Pressure-temperature curves on the three-dimensional pTx-phase boundary surfaces at constant compositions, x, were obtained (isopleths). The addition of NaCl shifts the fluid-fluid two-phase region to higher temperatures (ΔT) and pressures (ΔP) in the two systems (Fig. 6.6). The isobaric and isothermal shifts in certain regions could reach more than 100 K and 50 MPa with 8-wt% NaCl (Fig. 6.6a). Even with only 0.1-wt% NaCl, 15 K shift was observed near the respective CCs.

Near CP, water becomes a weakly-polar solvent. Sub- or super-critical water can form a homogeneous phase with non-polar gases (inert gases, H_2, O_2, N_2, CO_2), alkanes (up to C_{36}) and aromatics (naphthalene, benzene, alkyl-substituted

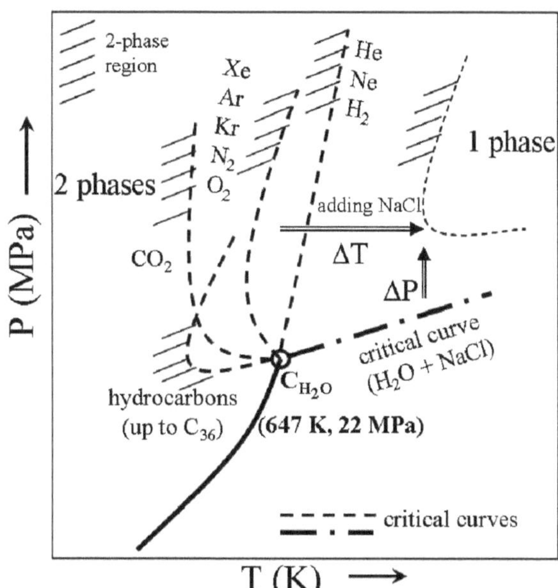

Fig. 6.7 Critical curves for binary and ternary aqueous systems

benzenes). The addition of salts (e.g., NaCl) significantly shifts the homogeneous phase region to higher temperatures and pressures (Fig. 6.7) [2–5, 10, 13, 15, 18]. Therefore, we can obtain fine particles by decreasing pressure and temperature of (or adding salts to) a supercritical organics solution.

6.2 Naphthalene (NT)

We produced fine particles from naphthalene (NT) by completely dissolving and subsequently precipitating it from SCW [19]. In Fig. 6.8, when {43 vol% NT + H_2O} with an air bubble (Fig. 6.8a) in DAC was heated at a rate of 1.1 K/s to 697 K and 315 MPa, air bubble disappeared at 491 K (Fig. 6.8b), at which density of water was calculated as 841 kg/m^3. The system was isochorely heated and pressure was calculated by knowing temperature and the density of water. At 627 K, NT particle became yellow (Fig. 6.8c) and changed to transparent at 661 K (Fig. 6.8d). The particle started to dissolve at 679 K (Fig. 6.8e). At maximum temperature of 697 K, all NT was dissolved in water (Fig. 6.8 g). After cooled, many particles precipitated along with a gas bubble (Fig. 6.8 h). The residue showed in Fig. 6.8 h was analyzed by FT-IR. Figure 6.9 showed little reaction occurred as compared the upper spectrum with that of standard naphthalene (the bottom curve).

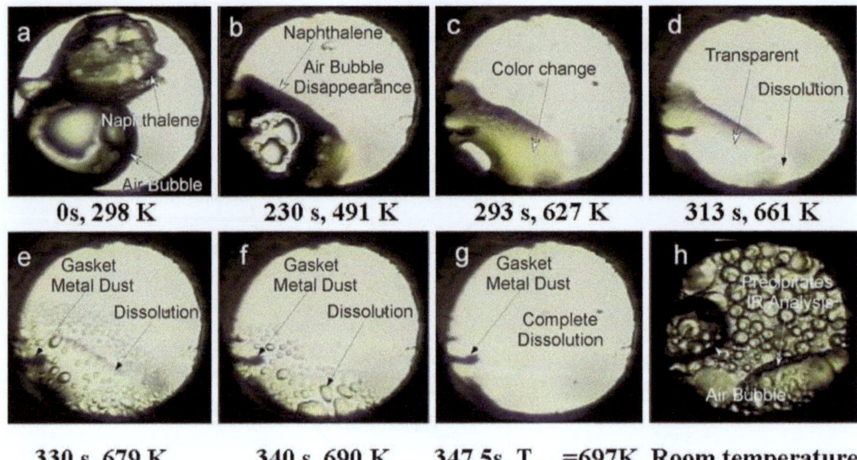

Fig. 6.8 Visual observation of {43-vol% NT + H_2O} heated to 697 and 315 MPa for the production of particles

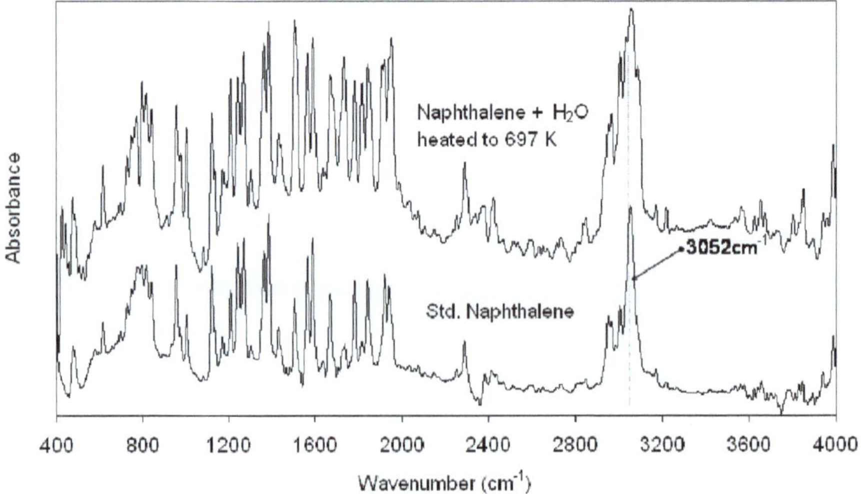

Fig. 6.9 IR spectra of the produced particles (*upper curve*) and standard naphthalene (*bottom curve*)

6.3 Polyethylene Terephthalate (PET)

Polyethylene terephthalate (PET) is a semi-crystalline thermoplastic polyester widely used in the manufacture of high strength fibers, photo-graphic films and soft-drink bottles. Recycling of PET has gained much attention. In our previous

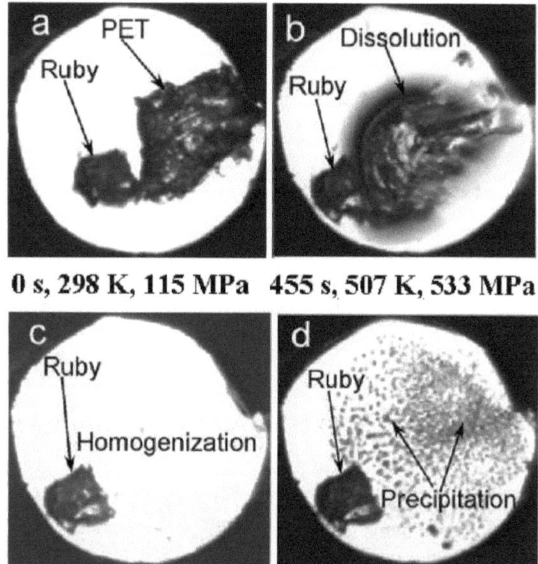

Fig. 6.10 Complete dissolution of PET in water at 516 K and production of fine particles by precipitation upon cooling (ruby: pressure sensor)

0 s, 298 K, 115 MPa   455 s, 507 K, 533 MPa

475s,T_{max}=516K,555MPa   486s, 398K, 283MPa

work in DAC [20], we completely dissolved PET, and subsequently fine particles were produced by precipitation upon cooling.

In Fig. 6.10, PET (25.3 vol.%) and water were heated slowly (0.5 K/s) from 298 K to a maximum temperature of 516 K and 555 MPa at an initial pressure of 115 MPa. PET particle started to dissolve at 507 K (Fig. 6.10b) after melting. Complete dissolution occurred at 516 K (Fig. 6.10c). The homogenous mixture was rapidly cooled and numerous fine particles precipitated as showed in Fig. 6.10d. Raman analysis of the precipitates showed the presence of PET as compared to the spectra of the standard PET and a typical particle label 1 in Fig. 6.11. Analysis also showed the presence of hydrolyzed PET due to an additional peak at 1644 cm^{-1} for particle 2 (Fig. 6.11). If heated further to higher temperatures up to supercritical region, terephthalic acid (TPA; monomer of PET) and PET oligomers-like crystals were obtained by precipitation of the supercritical solution. TPA microncrystals were also produced by solubilization and re-crystallization of amorphous TPA powders in SCW [20].

6.4 Nylon

Similar to the above methods, nylon particles were also produced [21]. In Fig. 6.12, nylon 6/6 (18.4 vol.%) and water were heated with a high rate (1.9 K/s) from 298 K to a maximum temperature of 613 K and 790 MPa at an initial pressure of 115 MPa. Nylon particle expanded and formed a film (Fig. 6.12b) after melting. Complete

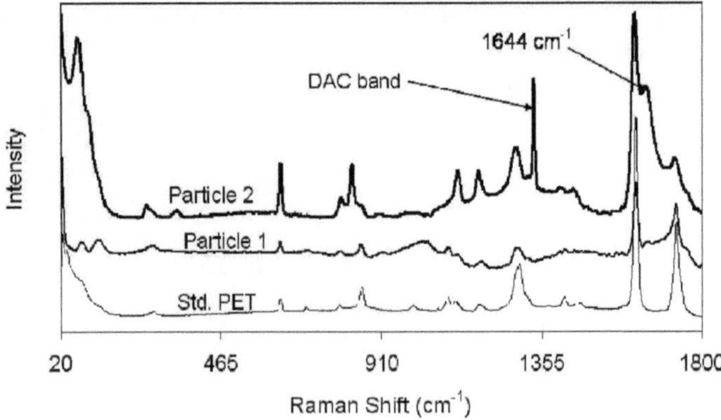

Fig. 6.11 Raman spectra of two selected particles from the precipitates of aqueous PET solution

Fig. 6.12 Complete
dissolution of nylon in water
at 613 K and production of
fine particles by precipitation
(ruby: pressure sensor)

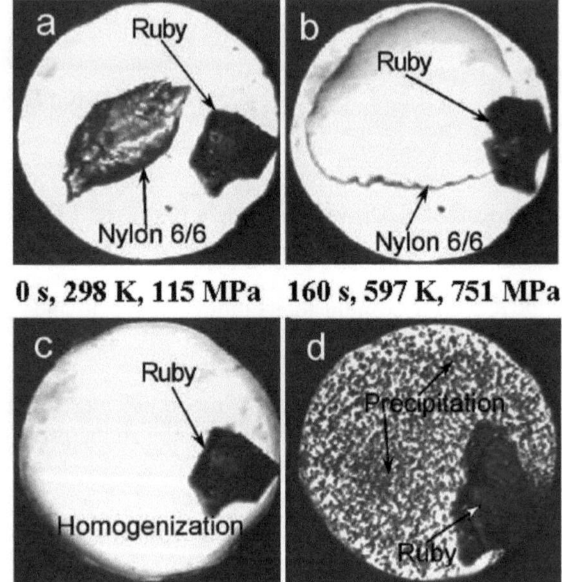

dissolution occurred at 613 K (Fig. 6.12c). The homogenous mixture was rapidly cooled and numerous fine particles precipitated (Fig. 6.12d). Raman analysis of the fine particles exhibited the presence of hydrolyzed nylon due to an additional peak at 2904 cm^{-1} for a randomly selected particle in Fig. 6.13. The results revealed nylon can completely dissolved in water, and fine particles were produced from the homogenous mixture by precipitation.

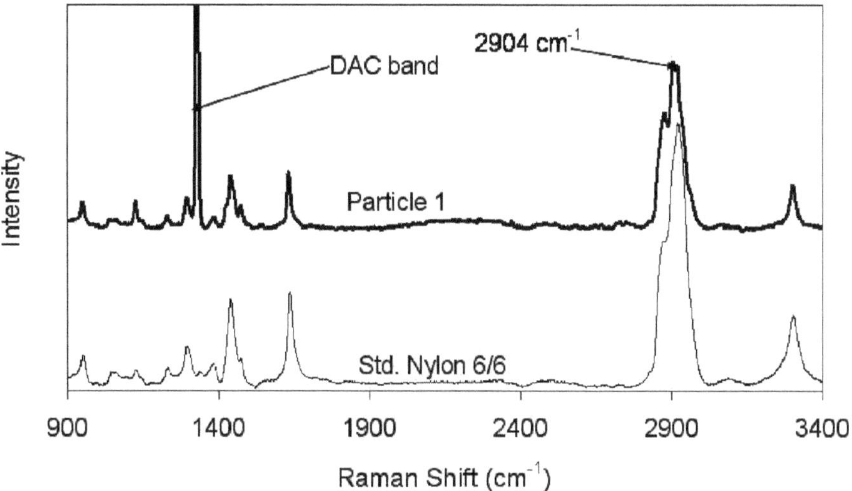

Fig. 6.13 Raman spectra of a selected particle from aqueous nylon solution by precipitation

6.5 Cellulose

In Fig. 6.14, microcrystalline cellulose I (27.5 vol.%) and water together with an air bubble (Fig. 6.14a) were heated rapidly (10 K/s) from 298 K to a maximum temperature of 602 K and 345 MPa. At 410 K (Fig. 6.14b), the air bubble disappeared and pressure was calculated as 0.33 MPa. The system was heated isochorely and cellulose started to dissolve at 597 K (Fig. 6.14c). Complete dissolution occurred at 602 K (Fig. 6.14d). After cooled, numerous cellulose particles were produced. By quick-quenching of the homogenous mixture of cellulose and water achieved by rapid-heating to 628~673 K (< 0.02~0.4 s) in a flow reactor, a new form of cellulose II was obtained continuously by reforming of the native cellulose I during precipitation [22].

If heated further to higher temperatures or holding longer reaction times, cellulose was undergoing homogeneously carbonization in SCW to produce numerous char particles as showed in Fig. 6.15.

6.6 Wood

Typical lignocellulosic biomass, wood and grass plants are composed of roughly 50% cellulose, 25% hemi-cellulose and 20% lignin, which are hydrolysable polymers built from sugar (glucosan, xylan, mannan, galactan) and phenylpropane units (lignin). There is no commercial production of fine wood particles whose size is smaller than 8 μm, and their processing techniques are very difficult for such fine

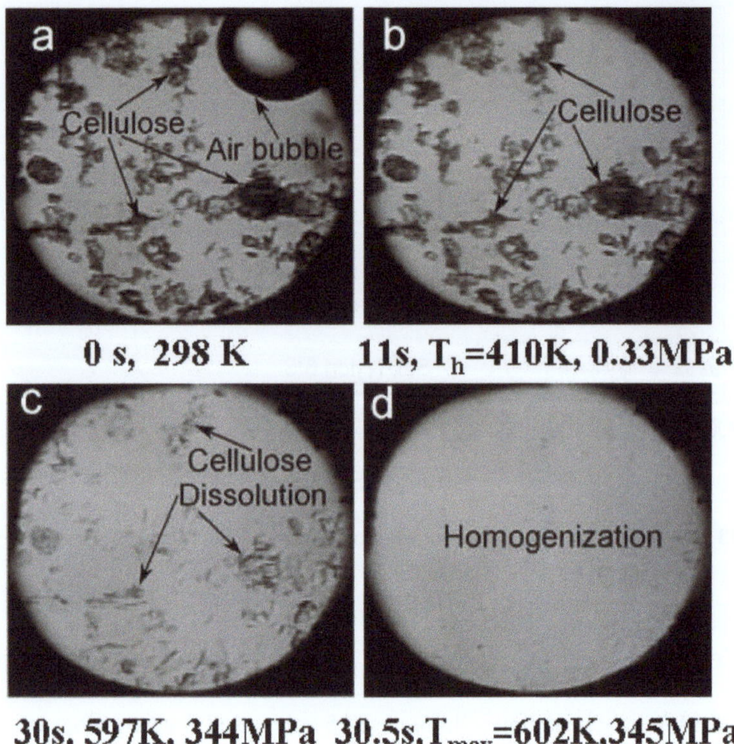

Fig. 6.14 Complete dissolution of cellulose in water at 602 K and 345 MPa (air bubble is for pressure calculation)

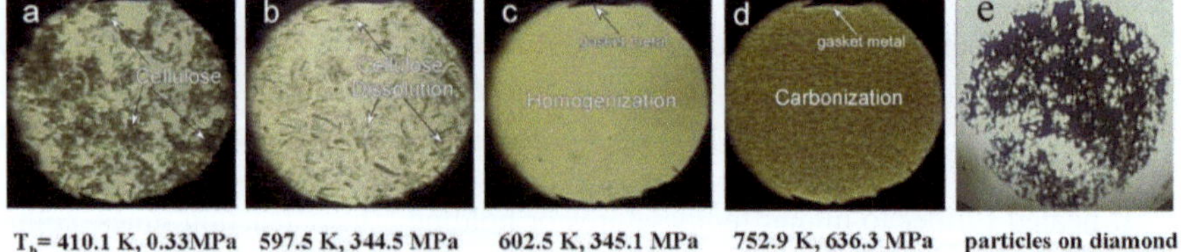

Fig. 6.15 Production of fine char particles by homogenous carbonization of cellulose in SCW

particles. Micron wood particles can find many applications in making composite materials, complicated woodcarvings, explosives, absorbents and natural wood coating. The mechanical and physical properties of synthetic wood particle materials can be easily changed by restructuring sub-micro particles. Recently, there was a breakthrough in our work that we could completely solubilize wood into wood solution using hot dilute Na_2CO_3 aqueous solution at temperature of 602~640 K and pressure of 14~106 MPa in less than 1.5 s [23−24]. By rapidly cooling the wood solution, micro-sized wood particles were obtained (Fig. 6.16).

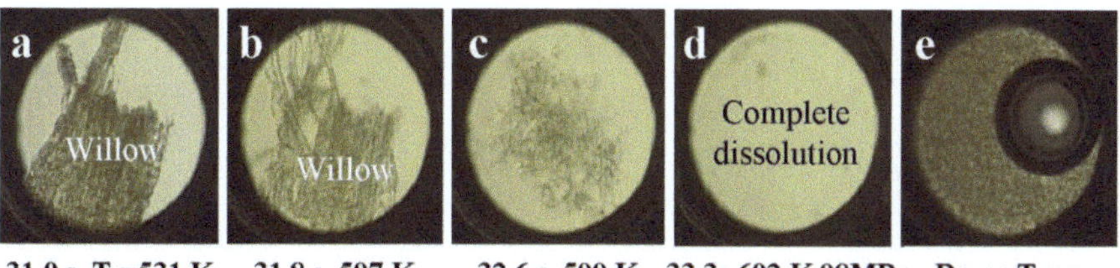

Fig. 6.16 Production of wood particles by complete dissolution and subsequent precipitation of willow in 0.8-wt%- Na_2CO_3 aqueous solution: heating {32-vol% willow + water + Na_2CO_3} to 602 K (at 9 K/s and $\rho = 787$ kg/m^3)

Fig. 6.17 SEM image of the micron particles in Fig. 6.16e

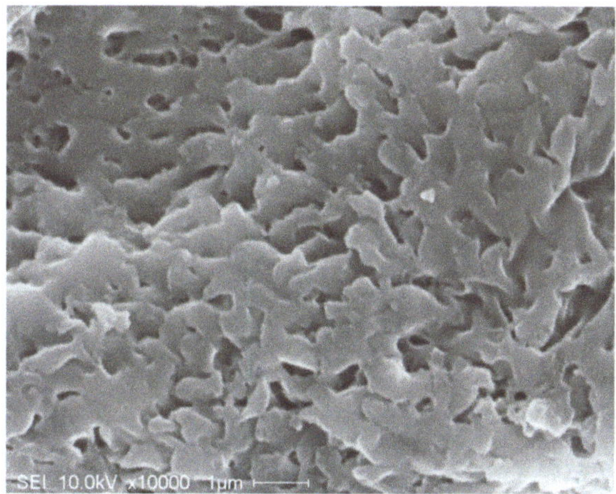

Willow was used as a model sample for wood in the following experiments. In Fig. 6.16, {32-vol.% willow + water + 0.8-wt% Na_2CO_3} was rapidly heated (9 K/s) to 602 K. At 531 K (Fig. 6.16a), gas bubbles disappeared and the water density was calculated as 787 kg/m^3. As the chamber was further isochoricly heated, willow particle shrank, and subsequently numerous tiny fibers unbound, dispersed and started to dissolved at 597 K (Fig. 6.16b). In just 1.5 s, all willow dissolved in water at 602 K and 98 MPa (Fig. 6.16b–d). After cooling, numerous particles were precipitated (Fig. 6.16e). SEM image (Fig. 6.17) showed that the particles were micron-sized (about 0.5 µm). The particles (Fig. 6.17) were decomposed slightly but with a strong willow character, because the broad willow characteristic peak at 2913 cm^{-1} still remained and a peak at 1740 cm^{-1} disappeared (Fig. 6.18; top curve vs. 1).

Many experiments were done in order to find the complete dissolution temperatures at different water densities (322~787 kg/m^3), willow concentrations (11~35 vol.%) and heating rates (7~16 K/s). At low water densities (< 550 kg/m^3), dissolution temperatures were at 609~638 K (or pressure of 14~20 MPa) for 11~28 vol.% willow concentrations at heating rates of 12~14 K/s in the L–V phase with

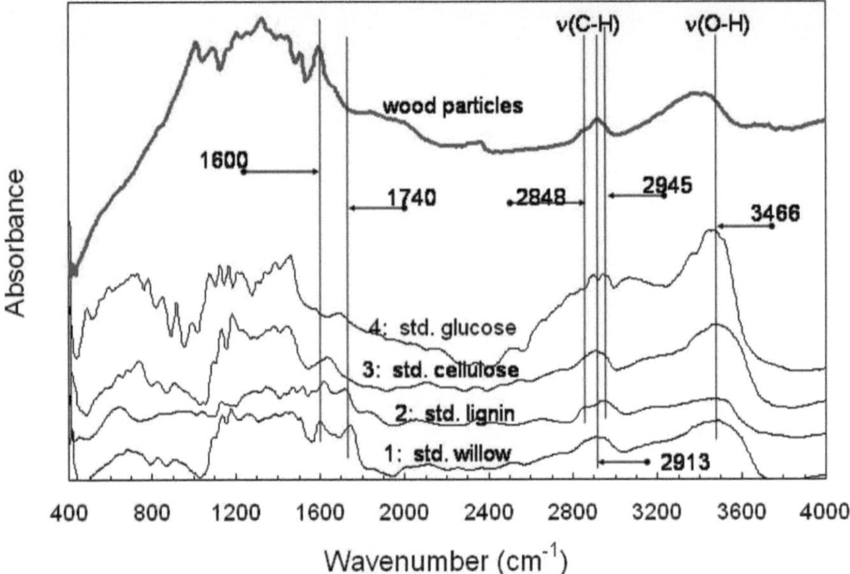

Fig. 6.18 IR spectra for the wood particles in Fig. 6.17 and standard biomass samples

gas bubbles. At higher water densities (570~790 kg/m^3), dissolution was undergoing in the liquid phase after gas bubbles disappeared, and dissolution temperature had a decreased trend as water density rose. At heating rates of 7~10 K/s, dissolution temperatures decreased from 628 to 603 K when water density increased from 640 to 790 kg/m^3 (33~98 MPa). At higher heating rates of 13~16 K/s, dissolution temperatures became higher but with the same trend that dropped from 640 to 623 K as water density rose from 640 to 790 kg/m^3 (17~106 MPa). Using the flow reactor in Figs. 2.9 and 2.10, micron-wood particles can be continuously produced by feeding wood slurry using a high pressure pump to the reactor.

6.7 Glucose

Fine particles can be produced by decomposition of aqueous glucose solution in SubCW and SCW [25]. In Fig. 6.19, aqueous glucose (1 M) with a gas bubble was slowly heated to 623 K at a heating rate of 0.19 K/s. At 473 K, the solution color changed to light yellow (Fig. 6.19b). The solution underwent further color changes from orange, red to dark-red color along with the precipitation of numerous particles at 513 K (Fig. 6.19c–e). More particles were precipitated and filled almost all the chamber as temperature increased further to 623 K (Fig. 6.19f–g). At 581 K, the gas bubble disappeared and water density was calculated as 695 kg/m^3. After reaction, numerous particles remained on the lower anvil (Fig. 6.19 h). IR results showed that the particle residue decomposed but still had a glucose-like character. The particles

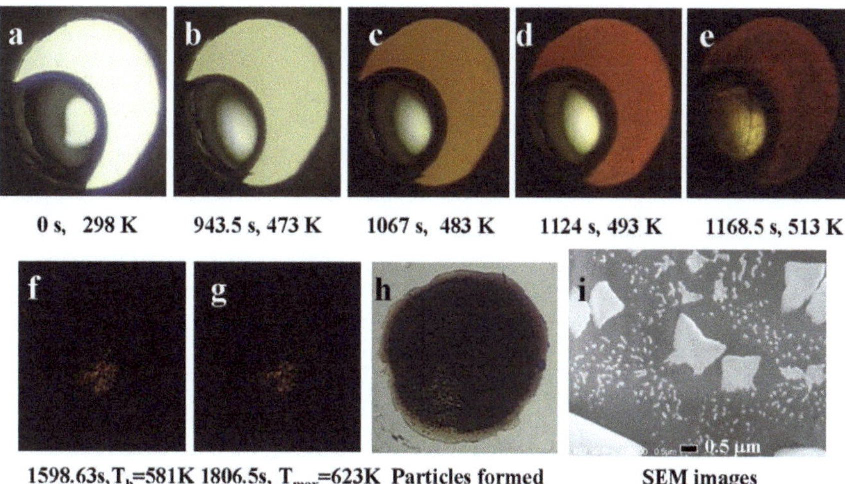

Fig. 6.19 Production of nanoparticles by slowly heating (0.19 K/s, $\rho = 695$ kg/m^3) aqueous glucose solution (1 M) to 623 K

were analyzed by SEM that showed they were sub-micron particles (Fig. 6.19i; about 100 nm).

Aqueous glucose was also studied during fast heating (4.6 K/s) to supercritical region at 683 K. After color changes from light yellow to dense yellow color at 673 K, numerous particles precipitated along with the appearance of an orange color at 683 K. Particles can be produced by slow heating at low temperatures (e.g., 623 K) and fast heating at higher temperatures (e.g., 683 K).

Since all the above hydrocarbons, polymers and biomass can be completely solubilized in SCW, their corresponding fine particles are easily produced continuously using the same flow reactor for oxide synthesis (Figs. 2.9 and 2.10) but a slurry pump is needed to feed water-insoluble biomass/polymer sample.

References

1. M.L. McGlashan, Phase equilibria in fluid mixtures. J. Chem. Thermodynamics **17**, 301–319 (1985)
2. E.U. Franck, Fluids at high pressures and temperatures. J. Chem. Thermodynamics **19**, 225–242 (1987)
3. M.L. Japas, E.U. Franck, High pressure phase equilibria and PVT-data of the water-oxygen system including water-air to 673 K and 250 MPa. Ber. Bunsenges. Physikal Chemie **89**, 1268–1275 (1985)
4. M.L. Japas, E.U. Franck, High pressure phase equilibria and PVT-data of the water-nitrogen system to 673 K and 250 MPa. Ber. Bunsenges. Physikal Chemie **89**, 793–800 (1985)
5. K. Todheide, E.U. Franck, Das zweiphasengebiet und die kritisehe kurve im system kohlendioxid-wasser bis zu drucken von 3500 bar. Z. Phys. Chem. Neue Folge **37**, 387–401 (1963)

6. E.U. Franck, H. Lentz, H. Welsch, The system water-xenon at high pressures and temperatures. Z. Phys. Chem. Neue Folge **93**, 95–108 (1974)
7. G. Wu, M. Heilig, H. Lentz, E.U. Franck, High pressure phase equilibria of the water-argon system. Ber. Bunsenges. Physikal Chemie **94**, 24–27 (1990)
8. A.E. Mather, R.J. Sadus, E.U. Franck, Phase equilibria in (water + krypton) at pressure from 31 MPa to 273 MPa and temperatures from 610 K to 660 K and in (water + neon) from 45 MPa to 255 MPa and from 660 K to 700 K. J. Chem. Thermodynamics **25**, 771–779 (1993)
9. N.G. Sretenskaja, R.J. Sadus, E.U. Franck, High-pressure phase equilibria and critical curve of the water + helium system to 200 MPa and 723 K. J. Phys. Chem. **99**, 4273–4277 (1995)
10. Y.S. Wei, R.J. Sadus, E.U. Franck, Binary mixtures of water + five noble gases: comparison of bimodal and critical curves at high pressures. Fluid Phase Equil. **123**, 1–15 (1996)
11. T.M. Seward, E.U. Franck, The system hydrogen – water up to 440 °C and 2500 bar pressure. Ber. Bunsenges. Physikal Chemie **85**, 2–7 (1981)
12. V.M. Shmonov, R.J. Sadus, E.U. Franck, High-pressure phase equilibria and supercritical pVT date of the binary water + methane mixture to 723 and 200 MPa. J. Phys. Chem. **97**, 9054–9059 (1993)
13. E. Brunner, Fluid mixtures at high pressures IX. Phase separation and critical phenomena in 23 (n-alkane + water) mixtures. J. Chem. Thermodynamics **22**, 335–353 (1990)
14. P.H. Van Konynenburg, R.L. Scott, Critical lines and phase equilibria in binary van der Waals mixtures. Phil. Trans. Roy. Soc. (London) A **298**, 495–540 (1980)
15. V.Z. Alwani, G.M. Schneider, Phasengleichgewichte, kritische erscheinungen und PVT-daten in binaren mischungen von wasser mit aromatischen kohlenwasserstoffen bis 420°C und 2200 bar. Ber. Bunsenges. Physikal Chemie **73**, 294–301 (1969)
16. G.M. Schneider, C.B. Kautz, D. Tuma, Physico-chemical principles of supercritical fluid science. NATO Sci. Ser. E: Appl. Sc. **366**(Supercritical Fluids), 31–68 (2000)
17. M. Gehrig, H. Lentz, E.U. Franck, The system water – carbon dioxide – sodium chloride to 773 K and 300 MPa. Ber. Bunsenges. Physikal Chemie **90**, 525–533 (1986)
18. T. Michelberger, E.U. Franck, Ternary systems water-alkane-sodium chloride and methanol-methane-sodium bromide to high pressures and temperatures. Ber. Bunsenges. Physikal Chemie **94**, 1134–1143 (1990)
19. S. Xu, Z. Fang, J.A. Kozinski, Decomposition of selected organic wastes during oxidation in supercritical water. Clean Air: Int. J. Energy Clean Environ. **5**(2), 39–52 (2004)
20. Z. Fang, R.L. Smith, Jr., H. Inomata, K. Arai, Phase behavior and reaction of polyethylene terephthalate-water systems at pressures up to 173 MPa and temperatures up to 490°C. J. Supercrit. Fluids **15**, 229–243 (1999)
21. R.L. Smith Jr., Z. Fang, H. Inomata, K. Arai, Phase behavior and reaction of nylon 6/6 in water at high temperatures and pressures. J. Appl. Polym. Sci. **76**(7), 1062–1073 (2000)
22. M. Sasaki, T. Adschiri, K. Arai, Production of cellulose II from native cellulose by near- and supercritical water solubilization. J. Agric. Food Chem. **51**, 5376–5381 (2003)
23. Z. Fang, C. Fang, Complete dissolution and hydrolysis of wood in hot water. AIChE J. **54**(10), 2751–2758 (2008)
24. Z. Fang, C. Fang, A method, equipment and applications for fast complete dissolution and hydrolysis of lignocellulosic biomass. US patent application#: 12671510 (2010); WO 2009/018709 A1 (2009); CN101235095A (2008)
25. Z. Fang, R.L. Smith, Jr., A method and equipment to produce dyes and nanoparticles using aqueous glucose. CN101265367A (2008)

Chapter 7
Conclusions and Future Prospects

7.1 Conclusion

In supercritical water, many nanocrystal oxides/oxide composites were produced rapidly (0.4 s~2 min) in a flow reactor or in a batch reactor for long time synthesis (e.g., 25 h) [1]. The SCW synthesis can be used for the production of ferrite magnetic pigments in recording media [Fe_3O_4, MFe_2O_4 (M = Co, Ni, Zn), $Ni_xCo_{1-x}Fe_2O_4$, $BaO{\cdot}6Fe_2O_3$], YAG: Tb phosphor for cathode ray tube screen, materials for lithium ion battery cathode ($LiCoO_2$, $LiMn_2O_4$), catalysts for car exhausts [e.g., $Ce_{1-x}Zr_xO_2$ (x = 0~1), $Zr_{1-x}In_xO_2$, $Zr_{1-x}Y_xO_2$], oxidation (La_2CuO_4) and gasification (ZrO_2, CeO_2, Ni), photo-catalysts (e.g., $K_2Ti_6O_{13}$, ZnO, TiO_2) for water decomposition, materials used in supporting of catalysts (e.g., boehmite; AlOOH) and materials (SnO_2, ZnO, In_2O_3) for electronics industry,. Organic capping ligands were successfully used to control and stabilize oxide particles (e.g., TiO_2, Fe_2O_3, Cu, CeO_2 nanocrystals). Nanostructured carbon film and fine metal oxide particles can be coated on the surface of materials in SCW. Inorganic salts crystals [e.g., $LiFePO_4$, $KCo_3Fe(PO_4)_3$ and $Mg_{3.5}H_2(PO_4)_3$] were also easily to be produced in SCW. The size, crystal structure and morphology of particles can be controlled by the feed concentrations (including molar ratio of initial samples), pH, types of salts, pressures, temperatures, heating rates (including preheating), organic modifications, reducing or oxidizing atmospheres (e.g., by adding H_2O_2, O_2, H_2) and reaction times in SCW. Hydrocarbons, biomass and polymer particles were also produced via solubilization and precipitation or carbonization in sub- and super-critical water. The fine particles produced from organics can be used in making composite materials or small sized bio-char that can be recycled to soil as fertilizer and to sequestrate carbon.

7.2 Future Prospects

Supercritical water synthesis in a flow reactor is a rapid, one-step and continuous method for getting nanocrystals of metal oxides, salts and other materials. But, the synthesis process is under severe conditions of high pressure (> 22 MPa) and high

Z. Fang, *Rapid Production of Micro- and Nano-particles Using Supercritical Water*,   87
Engineering Materials 30, DOI 10.1007/978-3-642-12987-2_7,
© Springer-Verlag Berlin Heidelberg 2010

ion concentration (K_w), and the technique is relative new {about 18 years [2, 3]}. Therefore, many physical and chemical data are not well-known and reaction mechanisms are not well clear. Few data are available for the phase behavior of metal salts and oxides. Few data are reported for the particle size and morphology of the ionic light metal salts that are precipitated in SCW. No experiments have been done to produce organic particles via solubilization and precipitation in continues flow reactors. Recently, Schubert et al. [4, 5] using a salt separator vessel in a continuously operated plant continuously separated many salts by precipitation in SCW. A versatile flow visualisation technique, light absorption imaging based on the absorption of light by dyed fluids, for quantifying mixing in a binary flow system has been developed [6]. The technique can be used for the fundamental research of the phase interaction or "*mixing mechanism*" in continuous reactors (Figs. 2.9 and 2.10).

We also successfully using the optical micro-reactor (DAC) to visually study the crystal formation and growing of erbium hydroxide in SubCW [7, 8], the cell can be used up to the supercritical region [9–11]. In batch reactors, various beautiful crystals (Fig. 7.1) were formed from {$Er(NO_3)_3 + NaOH + Na_2CO_3$} solution in SubCW at different pH values [8]. It was convenient to monitor the crystal growth and formation in-situ using DAC technique (Fig. 7.2). DAC is an idea tool for the study of SCW production of fine particles. Smith group [12, 13] also used the DAC for the visual study of Mn-doped zinc silicate formation (Fig. 7.3), and followed with in-situ synchrotron radiation XRD (SR-XRD) analysis (Fig. 7.4). Adschiri group has developed an experimental system to measure metal oxides solubility in SCW [14]. Lead (II) oxide and copper oxide solubility were measured from 523 to 773 K at 26~34 MPa. More recently, mechanism and kinetics of oxides formation in SCW were studied [15, 16].

Another disadvantage for SCW synthesis is the contamination by iron oxides and other metal oxides corroded from the metal reactor due to the acid- and base-like character of the supercritical state (Fig. 1.9) [17], particularly when acids and O_2 are presented [18] {e.g., HCl from PCBs [19–21]}. However, corrosion can be reduced

Fig. 7.1 SEM micrographs of different crystal shapes formed in batch reactors at 573 K and 25 MPa from {$Er(NO_3)_3 + NaOH + Na_2CO_3$}: (**a**) pH = 12 and (**b**) pH = 6

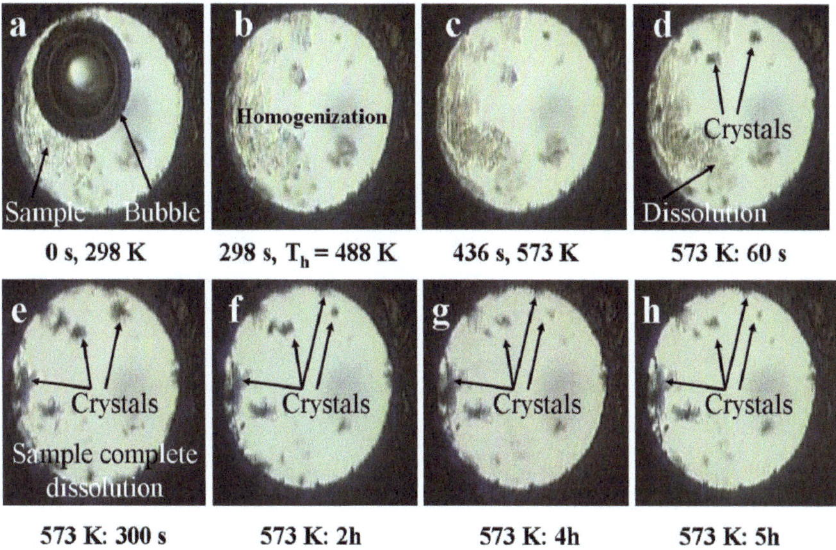

Fig. 7.2   Visual observation of crystal formation in DAC during heating {[Er(NO$_3$)$_3$ + NaOH + Na$_2$CO$_3$]; pH = 12} to supercritical region at 573 K and 133 MPa at a heating rate of 0.6 K/s and held for 5 h (ρ = 847 kg/m^3)

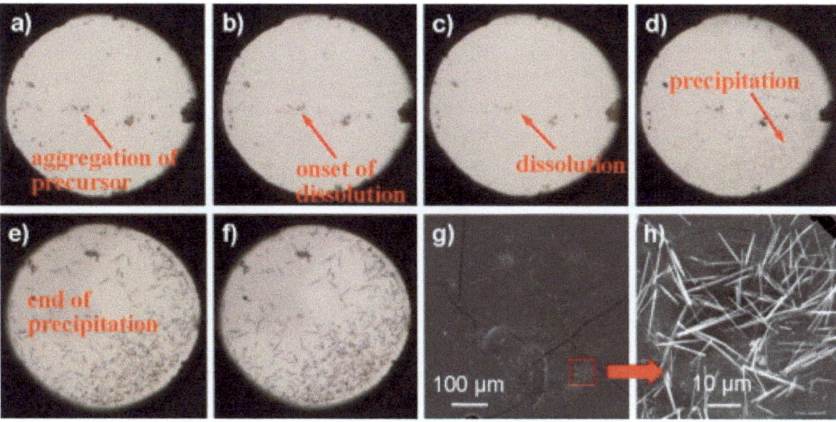

Fig. 7.3   Visual Observation of formation of crystals of Mn-doped zinc silicate in water in DAC at the average heating rate of 13.2 K/s: (**a**) 303 K, 3.1 MPa, (**b**) 476 K, 295 MPa, (**c**) 567 K, 494 MPa, (**d**) 669 K, 718 MPa, (**e**) 923 K, 1,250 MPa, (**f**) after cooling, and SEM images of Mn-doped zinc silicate produced, (**g**) overall (cracked background is carbon tape) and (**h**) close-up at 110× magnification. Reprinted with permission from [12]. Copyright © 2007, Elsevier

by using corrosion-resistant alloys (e.g., high Ni-based G-30), liners & coating {e.g., monolithic alumina ceramics [22], noble metals of Pt and Rh [23]}, adding ionic light metal salts to form a fine particle layer [24] and modifying reactor design [25].

Fig. 7.4 One-dimensional in-situ SR-XRD spectra of products forming in water in DAC: **(a)** 611 K (0.59 GPa, 341 s), **(b)** 630 K (0.63 GPa, 372 s), **(c)** 647 K (0.67 GPa, 399 s, **(d)** 667 K (0.71 GPa, 427 s), and **(e)** 760 K (0.91 GPa, 566 s). Reprinted with permission from [13]. Copyright © 2009, Elsevier

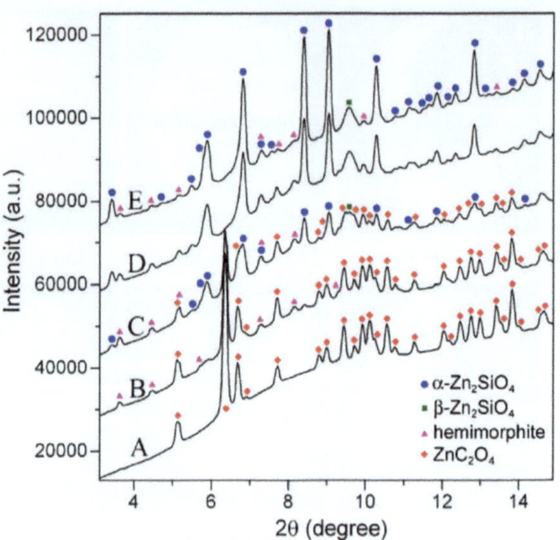

Fig. 7.5 Schematic of the "nozzle" reactor design with ideal heating/cooling profile. Reprinted with permission from [27]. Copyright © 2006, Elsevier

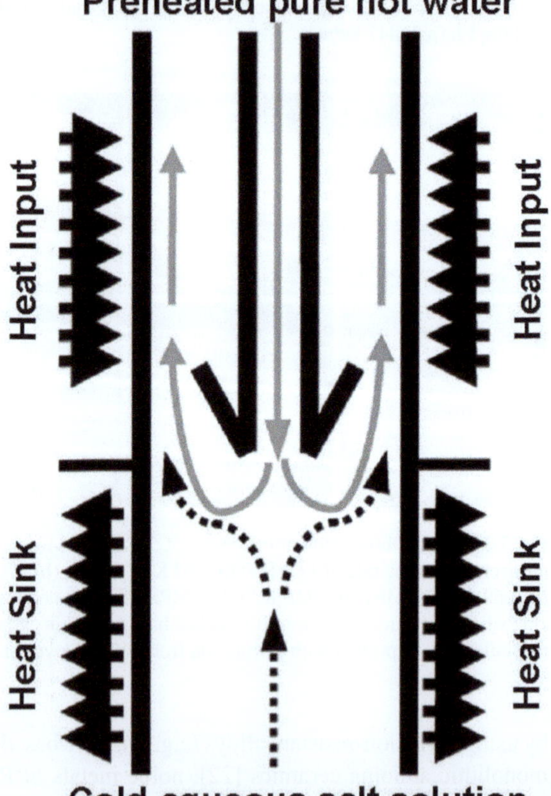

Supercritical hydrothermal synthesis is a relatively simple and environmentally friendly process for the production of potentially valuable metal oxide, salt particles and organic particles. However, its continuous process has not yet found industrial applications to date due to poor process reliability, reproducibility and control in flow systems.

Mixing systems are important to decide particle size, size distribution and crystallinity. The particles synthesized with the T-piece geometry are smaller with a narrower size distribution, possibly indicating a more effective mixing, than particles synthesized at the same conditions with concentric counter-flow geometry [26].

New engineering design for the flow process is needed, for example, at the mixing point in Fig. 2.9, an optimized *"nozzle"* reactor (Fig. 7.5) was developed and resulted in a dramatic improvement in process reproducibility and reliability [27]. A new *"swirling micro mixer"* for continuous hydrothermal synthesis was also developed and demonstrated reduction of particle size, aggregation and dispersibility [28]. In a practical experiment of boehmite (AlOOH) fine particle synthesis, the developed *"micro mixer"* demonstrated average particle sizing down from 100~200 nm to 60 nm.

References

1. R.B. Yahya, H. Hayashi, T. Nagase, T. Ebina, Y. Onodera, N. Saitoh, Hydrothermal synthesis of potassium hexatitanates under subcritical and supercritical water conditions and its application in photocatalysis. Chem. Mater. **13**, 842–847 (2001)
2. T. Adschiri, K. Kanazawa, K. Arai, Rapid and continuous hydrothermal synthesis of boehmite particles in subcritical and supercritical water. J. Am. Ceram. Soc. **75**, 2615–2618 (1992)
3. T. Adschiri, K. Kanazawa, K. Arai, Rapid and continuous hydrothermal crystllization of metal oxides particles in supercritical water. J. Am. Ceram. Soc. **75**, 1019–1022 (1992)
4. M. Schubert, J.W. Regler, F. Vogel, Continuous salt precipitation and separation from supercritical water. Part 1: Type 1 salts. J. Supercrit. Fluids **52**, 99–112 (2010)
5. M. Schubert, J.W. Regler, F. Vogel, Continuous salt precipitation and separation from supercritical water. Part 2. Type 2 salts and mixtures of two salts. J. Supercrit. Fluids **52**, 113–124 (2010)
6. P.J. Blood, J.P. Denyer, B.J. Azzopardi, M. Poliakoff, E. Lester, A versatile flow visualisation technique for quantifying mixing in a binary system: application to continuous supercritical water hydrothermal synthesis (SWHS). Chem. Eng. Sci. **59**(14), 2853–2861 (2004)
7. H. Assaaoudi, Z. Fang, J.E. Barralet, A.J. Wright, I.S. Butler, J.A. Kozinski, Synthesis, characterization and properties of erbium-based nanofibers and nanorods. Nanotechnology **18**, 445606 (7pp) (2007)
8. H. Assaaoudi, Z. Fang, I.S. Butler, J.A. Kozinski, Synthesis of erbium hydroxide microflowers and nanostructures in subcritical water. Nanotechnology **19**, 185606 (8 pp) (2008)
9. W.A. Bassett, A.H. Shen, M. Bucknum, I.M. Chou, A new diamond-anvil cell for hydrothermal studies to 2.5 GPa and from –190°C to 1200°C. Rev. Sci. Instrum. **64**, 2340–2345 (1993)
10. R.L. Smith Jr., Z. Fang, Techniques, applications and future prospects of diamond anvil cells for studying supercritical water systems. J. Supercrit. Fluids **47**, 431–446 (2009)
11. T. Wang, R.L. Smith Jr., H. Inomata, K. Arai, Reactive phase behavior of aluminum nitrate in high temperature and supercritical water. Hydrometallurgy **65**(2–3), 159–175 (2002)

12. M. Takesue, K. Shimoyama, S. Murakami, Y. Hakuta, H. Hayashi, R.L. Smith, Phase formation of Mn-doped zinc silicate in water at high-temperatures and high-pressures. J. Supercrit. Fluids **43**(2), 214–221 (2007)
13. M. Takesue, K. Shimoyama, K. Shibuki, A. Suino, Y. Hakuta, H. Hayashi, Y. Ohishi, R.L. Smith, Formation of zinc silicate in supercritical water followed with in situ synchrotron radiation X-ray diffraction. J. Supercrit. Fluids **49**(3), 351–355 (2009)
14. K. Sue, Y. Hakuta, R.L. Smith Jr., T. Adschiri,, K. Arai, Solubility of lead(II) oxide and copper(II) oxide in subcritical and supercritical water. J. Chem. Eng. Data, **44**, 1422–1426 (1999)
15. M.N. Danchevskaya, Y.D. Ivakin, S.N. Torbin, G.P. Muravieva, The role of water fluid in the formation of fine-crystalline oxide structure. J. Supercrit. Fluids **42**(3), 419–424 (2007)
16. J.H. Lee, J.Y. Ham, Mechanisms for metal oxide particles synthesized in Supercritical water. J. Ind. Eng. Chem. **13**(5), 835–841 (2007)
17. M. Uematsu, E.U. Franck, Static dielectric constant of water and steam. J. Phys. Chem. Ref. Data **9**(6), 1291–1306 (1980)
18. L.B. Kriksunov, D.D. Macdonald, Corrosion in supercritical water oxidation systems: A phenomenological analysis. J. Electrochem. Soc. **142**(12), 4069–4073 (1995)
19. Z. Fang, S. Xu, I.S. Butler, R.L. Smith Jr., J.A. Kozinski, Destruction of decachlorobiphenyl using supercritical water oxidation. Energy & Fuels **18**(5), 1257–1265 (2004)
20. Z. Fang, S.K. Xu, J.A. Kozinski, Flameless oxidation of chlorinated wastes in supercritical water using sodium carbonate as a stimulant. Proc. Combust. Inst. **29**(2), 2485–2492 (2002)
21. Z. Fang, S.K. Xu, R.L. Smith Jr., J.A. Kozinski, K. Arai, Destruction of deca-chlorobiphenyl in supercritical water under oxidizing conditions with and without Na_2CO_3. J. Supercrit. Fluids **33**(3), 247–258 (2005)
22. N. Boukis, N. Claussen, K. Ebert, R. Janssen, M. Schacht, Corrosion screening tests of high-performance ceramics in supercritical water containing oxygen and hydrochloric acid. J. Eur. Ceram. Soc. **17**, 71–76 (1997)
23. K.W. Downey, R.H. Snow, D.A. Hazlebeck, A.J. Roberts, Corrosion and chemical agent destruction: Research on supercritical water oxidation of hazardous military wastes. ACS Symp. Ser. **608**, 313–326 (1995)
24. P. Muthukumaran, R.B. Gupta, Sodium-carbonate-assisted supercritical water oxidation of chlorinated waste. Ind. Eng. Chem. Res. **39**, 4555–4563 (2000)
25. D.B. Mitton, J.H. Yoon, J.A. Cline, H.S. Kim, N. Eliaz, R.M. Latanision, Corrosion behavior of nickel-based alloys in supercritical water oxidation systems. Ind. Eng. Chem. Res. **39**(12), 4689–4696 (2000)
26. L.L. Toft, D.F. Aarup, M. Bremholm, P. Hald, B.B. Iversen, Comparison of T-piece and concentric mixing systems for continuous flow synthesis of anatase nanoparticles in supercritical isopropanol/water. J. Solid State Chem. **182**(3), 491–495 (2009)
27. E. Lester, P. Blood, J. Denyer, D. Giddings, B. Azzopardi, M. Poliakoff, Reaction engineering: The supercritical water hydrothermal synthesis of nano-particles. J. Supercrit. Fluids **37**(2), 209–214 (2006)
28. Y. Wakashima, A. Suzuki, S. Kawasaki, K. Matsui, Y. Hakuta, Development of a new swirling micro mixer for continuous hydrothermal synthesis of nano-size particles. J. Chem. Eng. Jpn. **40**(8), 622–629 (2007)

MIX
Papier aus verantwortungsvollen Quellen
Paper from responsible sources
FSC® C105338
www.fsc.org

If you have any concerns about our products,
you can contact us on
ProductSafety@springernature.com

In case Publisher is established outside the EU,
the EU authorized representative is:
Springer Nature Customer Service Center GmbH
Europaplatz 3, 69115 Heidelberg, Germany

Printed by Libri Plureos GmbH
in Hamburg, Germany